GERMAN DIVISIONS IN WORLD WAR I

Volume 4: 14th – 27th Divisions

By Dirk Rottgardt

THE NAFZIGER COLLECTION
2007

Printed in the United States of America

ISBN 978-1-58545-168-1

We are interested in hearing from authors with book ideas.

Published by The Nafziger Collection, Inc.
PO Box 1522, West Chester, OH 45071-1522

E-mail: Nafziger@fuse.net
On-line Catalog: http://home.fuse.net/nafziger

German 28th-55th Divisions

28th Infantry Division: The division belonged to the Baden Contingent within the Prussian Army, the 14th Army Corps, and started the war in Upper Alsace, fighting at Muelhausen/Mulhouse on 9 August and at Altkirch on the 13th defending German Alsace against the French 1914 invasion. In early September it fought at Mortange and then moved west of Pont à Mousson where it was engaged 20 to 29 September. On 4 October it was shifted to the Artois. From October 1914 to May 1915 the division slowly exhausted its regiments in fighting on the Notre Dame de Lorette plateau. On 30 November 1914, the 110th Grenadier Regiment reported that it had suffered 58 officers and 3,814 men as casualties since the beginning of the campaign. In August 1914 the division contained:

28th Infantry Division:
- 55th Infantry Brigade:
  - 109th Leibgrenadier Regiment[1]
  - 110th Grenadier Regiment
- 56th Infantry Brigade:
  - 40th Fusilier Regiment[2]
  - 111th Infantry Regiment
- Cavalry:
  - 1/ to 4/5th Jäger zu Pferd Regiment[3]

28th Field Artillery Brigade:
- 14th Field Artillery Regiment
- 50th Field Artillery Regiment

Support Troops:
- 2/, 3/14th Pioneer Battalion
- 28th Divisional Bridging Train
- 2nd Medical Company

The division suffered heavily during the winter 1914/1915, especially the 110th Grenadiers and 40th Fusiliers. Between 9 April and 3 May it suffered again heavily at Carency, Ouvrages and Blancs in the Artois. In the course of these three battles the 111th Infantry was almost completely annihilated. On 10 May the 1st Bn/111th Infantry had only 3 officers and 272 men left. Its casualty lists show 32 officers and 1,737 men. On 15 May the division was relieved and reorganized. On 25 May it was returned to the lines and again suffered badly.

On 13 June the division left the Artois and moved to the Champagne, occupying a quiet sector on the front between Bétheny and the Sillery-Beine road. In the September 1915 offensive, the division detached one battalion, each, of the 109th and 110th Grenadiers as reinforcements in the Somme-Py area. On 19/20 October the division carried out a gas attack with rather poor results as it was not accustomed to those new

---

1 Leibgrenadier Regiment means (Grand Duke's) Own Grenadier Regiment. This regiment was the Guards of the Grand Duke of Baden.

2 This was not a Baden Regiment, but a "truly" Prussian one recruited in Hohenzollern-Sigmaringen, the original home of the Hohenzollern dynasty, and re-incorporated into the Prussian state in the early 19th century.

3 Mobilized in August 1914, leaving the 5th Sqn as Ersatz Sqn at Offenburg, then at Karlsruhe and finally at Gochenheim, 14th Corps District. The 1/, 3/ and 4/5th Jäger zu Pferd Regiment already in December 1914 joined the improvised Lippe Cavalry Division of the 3rd Army (see note on 12th Infantry Division). The 2nd stayed with the 28th Infantry Division. When the Lippe Division was dissolved in January 1916, the 1st Sqn joined the 3rd Line-of-Communication Inspection, in July 1916 the 185th Infantry Division and subsequently the 58th Infantry Division, the 24th Reserve Division and finally, still in 1916, the 8th Landwehr Division. The 3rd also was at first with the 3rd Line-of Communication Inspection and then stayed from July 1916 onwards with the 185th Infantry Division while the 4th joined the 29th Infantry Division.

tactics. It was relieved around 10 November and was sent to the Tahure-Butte du Mesnil area where it stayed during the winter. In 1915 the division contained:

28th Infantry Division:
  55th Infantry Brigade:
    109th Leibgrenadier Regiment
    110th Grenadier Regiment
  56th Infantry Brigade:
    40th Fusilier Regiment[4]
    111th Infantry Regiment
  Cavalry:
    2/5th Jäger zu Pferd Regiment
28th Field Artillery Brigade:
  14th Field Artillery Regiment
  50th Field Artillery Regiment
Engineers:
  ? Company, 14th Pioneer Battalion
  28th Divisional Bridging Train
Signals:
  28th Telephone Detachment

At the end of April the division was moved again. After a week of rest near Vouziers, it went into the lines about 5 May near Maisons de Champagne. During this time it received intensive training. In July the division, without the109th Leibgrenadier Regiment, which stayed in the old sector, was gradually sent from the Champagne to the Somme. Between 16 and 20 July it took up positions between the Somme and Barleux. It remained there until early October, suffering heavy losses. In early October it returned to the Champagne. It remained there until January 1917. In 1916 the division contained:

28th Infantry Division:
  55th Infantry Brigade:
    109th Leibgrenadier Regiment
    110th Grenadier Regiment
  56th Infantry Brigade[5]:
    40th Fusilier Regiment
    111th Infantry Regiment
  Cavalry:
    2/5th Jäger zu Pferd Regiment[6]
  28th Field Artillery Brigade:
    14th Field Artillery Regiment
    50th Field Artillery Regiment
Engineers[7]:
  1/14th Pioneer Battalion
  3/14th Pioneer Battalion
  28th Trench Mortar Company[8]
  28th Divisional Bridging Train
Signals:
  28th Telephone Detachment
Attached:
  29th Anti Aircraft Section
  1st Bavarian Labor Battalion

In January 1917 the division was sent to the Verdun area. It would remain there until early September, taking part in the mid-August attacks. On 6 April 1917 the 111th Infantry Regiment was sent to the rebuilding 56th Division. In mid-September the division went to Alsace. In late October it left for the Montmedy area, then Laon, and finally Cambrai where it was engaged on 30 November. In December it rested in the Ardennes and in February 1918 it went into the Mont Cornillet sector. Even if the

---

[4] A Prussian regiment stationed in peace at Rastatt, Baden, but recruited in the Prussian exclave of Hohenzollern.

[5] Joined on 30 November 1916 the 52nd Reserve Division and, on 4 April 1917, the 54th Infantry Division.

[6] From 31 July to 17 August 1916 there was a combined Cavalry Regiment, 28th Infantry Division consisting of the Staff/, 2/, 4/16th Dragoons and the 2/5th Jäger zu Pferd Regiment.

[7] *Stein* says that the 9/28th Pioneer Battalion, for some unspecified time, perhaps in 1915/16, also served with this division. It would subsequently go to the 18th Army, the 28th Reserve Division, the 15th Reserve Division and finally the 103rd Infantry Division.

[8] Raised in October 1915 from the 51st Heavy, the 183rd Medium and the 227th Light Trench Mortar Detachments. Distributed in autumn 1918 among the 40th Fusilier and the 109th and 110th Grenadier Regiments.

division, with the exception of the Prussian 40th Fusilier Regiment, was recruited in Baden there were by 1917some drafts from the 4th Corps District. In 1917 the division contained:

28th Infantry Division:
- 56th Infantry Brigade:
  - 40th Fusilier Regiment
  - 109th Leibgrenadier Regiment
  - 110th Grenadier Regiment
- Cavalry:
  - 2/5th Jäger zu Pferd Regiment
- 28th Artillery Command:
  - 14th Field Artillery Regiment
- 130th Pioneer Battalion:
  - 2/, 3/14th Pioneer Battalion
  - 28th Trench Mortar Company

Signals:
- 28th Telephone Detachment

Medical:
- 35th Medical Company
- 262nd Field Hospital
- 292nd Field Hospital
- ? Veterinary Hospital

Transport:
- 67th Truck Train
- ? Light Munitions Column

Attached:
- 40th Machine Gun Sharpshooter Battalion

From January 1918 to mid-May the division remained in the Butte du Mesnil Region. On 23 May the division moved to the Laon Region and went into the lines to reinforce the Aisne battle until 8 June. From 13 June to 1 July it fought in the Second Battle of the Marne. On 24 July it returned to the battle near Chaumuzy, southwest of Rheims. From late August to 14 September, it was at rest near Baroucourt. In mid-September until the end of the war it continued in the lines in the Woevre sector, fighting until the signing of the armistice. In 1918 the division contained:

28th Infantry Division:
- 56th Infantry Brigade:
  - 40th Fusilier Regiment
  - 109th Leibgrenadier Regiment
  - 110th Grenadier Regiment
- Cavalry:
  - 2/5th Jäger zu Pferd Regiment
- 28th Artillery Command:
  - 14th Field Artillery Regiment
  - 55th Foot Artillery Battalion
  - 801st Light Ammunition Column
  - 991st Light Ammunition Column
  - 1129th Light Ammunition Column

14th Pioneer Battalion:
- 2/, 3/14th Pioneer Battalion
- 94th Searchlight Section

28th Signals Command:
- 28th Telephone Detachment
- 55th Wireless Detachment

Medical:
- 35th Medical Company
- 262nd Field Hospital
- 262nd Field Hospital
- 28th Veterinary Hospital

Transport:
- 561st Motor Transport Column

28th Reserve Division: The 28th Reserve Division was formed in Baden in the 14th Corps District. It was engaged in the Bruche Valley on 15 August, fighting towards Meurthe until 5 September, the 111th Reserve suffering two-thirds of its strength as casualties. After 5 September the division went into the Somme and ended up near Cambrai by 27 September. In August 1914 the division contained:

28th Reserve Division[9]:
- 55th Reserve Brigade:
  - 40th Reserve Regiment[10]
  - 109th Reserve Regiment[11]
  - 8th Reserve Jäger Battalion[12]
- 56th Reserve Brigade:
  - 110th Reserve Regiment[13]
  - 111th Reserve Regiment[14]
  - 14th Reserve Jäger Battalion[15]

Cavalry:
- 1/, 2/, 3/8th Reserve Dragoon Regiment[16]

Artillery:
- 29th Reserve Field Artillery Regiment (9 btrys)[17]

Engineers:
- 1st & 2nd Reserve Companies, 13th Pioneer Battalion
- 28th Reserve Divisional Bridging Train
- 14th Reserve Medical Company

The division occupied the sector from Ovillers-Fricourt until July 1916. In April 1915 the 40th Reserve Regiment was sent to the 115th Infantry Division and the 55th Reserve Brigade was disbanded. The 8th and 14th Reserve Jäger Battalions were also detached – one in January and the second in May. Elements of the division were in reserve in the St. Quentin Region during August and September. In 1915 the division contained:

28th Reserve Division:
- 55th Reserve Brigade[18]:
  - 109th Reserve Regiment
  - 110th Reserve Regiment
- 56th Reserve Brigade:
  - 40th Reserve Regiment
  - 111th Reserve Regiment
  - 9th Reserve Jäger Battalion
  - 14th Reserve Jäger Battalion
  - 55th Landwehr Brigade Ersatz Bn[19]

Cavalry:
- 8th Reserve Dragoon Regiment

Artillery:
- 29th Reserve Field Artillery Regiment (9 btrys)

Engineers:
- 1st & 2nd Reserve Companies, 13th Pioneer Battalion
- 28th Reserve Divisional Bridging Train

Signals:
- 28th Reserve Telephone Detachment

---

[9] No Baden reserve troops were the Rhinish 8th Reserve Jäger Battalion and the Mecklenburg 14th Reserve Jäger Battalion.

[10] Raised in August 1914 in the 14th Corps District at Mannheim and Heidelberg (2nd Bn). It recurited Baden, not Hohenzollern personnel. No machine guns. Disbanded in September 1918 and distributed among the 109th, 110th and 111th Reserve Regiments.

[11] Raised in August 1914 in the 14th Corps District at Karlsruhe and Bruchsal (2nd Bn).

[12] Raised in August 1914 in the 15th Corps District at Schlettstadt/Selestat and Neubreisach/Neuf Brisach by the Rhinish 8th Jäger Battalion. Joined in October 1917 the 13th Jäger Regiment.

[13] Raised in August 1914 in the 14th Corps District at Rastatt (Staff, 1st Bn), Karlsruhe (2nd Bn) and Mannheim (3rd Bn).

[14] Raised in August 1914 in the 14th Corps District at Donaüschingen (1st Bn), Stockach (2nd Bn) and Constance (Staff, 3rd Bn).

[15] Raised in August 1914 in the 15th Corps District at Colmar and Neubreisach/Neuf Brisach in Alsace by the Mecklenburg 14th Jäger Battalion with Mecklenburg reservists. Joined in May 1915 the 2nd Jäger Regiment of the Alpine Corps.

[16] Raised in August 1914 by the 21st Dragoon Regiment with three squadrons. Sqns seconded from the second half of 1916 onwards to various military governments in Poland.

[17] The Baden 29th Reserve Field Artillery Regiment had a number differing from that if its division. *Cron* claims that the regiment had three battalions, while the other Prussian reserve field artillery regiments just had two. Instead, *Kling* tells of only two battalions. Raised, according to him, on mobilization in August 1914 by the 14th and 65th Field Artillery Regiments at Karlsruhe and Ludwigsburg. The 2nd Battalion was a Württemberg formation!

[18] Disbanded on 5 April 1915.

[19] Raised in the 14th Corps District from two companies, each, of the Ersatz battalions of the 109th Landwehr and 109th Reserve Regiments. Became 3rd Bn/94th Reserve Regiment in 1916.

In 1916 the *Histories* reports that the 29th Reserve Field Artillery Regiment had "29 btrys," but this is must be a typographical error. Between October 1915 and July 1916 in the Somme Region the division had few losses. On 1 July 1916 it suffered the entire weight of the British offensive north of the Somme and had very heavy casualties. The 111th Reserve Regiment alone lost 39 officers and 1,821 men. On 4 July the division was withdrawn and rested. On 10 July it went to a quiet sector in the Champagne. On 5 October the division went back to the Somme and was in action at the end of October losing heavily. On 28 October it was sent to the Stenay area and reorganized. In early November the division went to the Verdun Region. In 1916 the division contained:

28th Reserve Division:
  56th Reserve Brigade:
    109th Reserve Regiment
    110th Reserve Regiment
    111th Reserve Regiment
  Cavalry:
    8th Reserve Dragoon Regiment
    3/22nd Dragoon Regiment[20]
  28th Reserve Field Artillery Brigade:
    28th Reserve Field Artillery Regiment (6 btrys)[21]
    29th Reserve Field Artillery Regt (6 btrys)
    1st Ersatz Bn, 76th Field Artillery Regt
  Engineers[22]:
    1st & 2nd Reserve Companies, 13th Pioneer Bn[23]
    228th Trench Mortar Company[24]
    28th Reserve Divisional Bridging Train
  Signals:
    28th Reserve Telephone Detachment

The division remained in the Avecourt sector in the Verdun Region until some time between 7-15 April. It went to the Californie Plateau and was attacked by the French on 4 May, again suffering heavy losses. Its subsequent counterattack also cost it heavily. Those were made up by 1,100 recruits and men from the disbanded 626th Infantry Regiment. The division was sent to Verdun again until 20 May. Suffering from dysentery, the division was withdrawn on 8 July and rested until early August. It then went into the lines on the right bank of the Meuse and suffered heavy losses from a French attack on 20 August – 47 officers and 1,150 men as prisoners. It was relieved on 30 August. In September it was sent to the Champagne. In addition to Baden personnel, the division had by 1917 also taken in replacements from the Rhenish Hesse – 11th Corps District –, the Rhine Province – 8th Corps District – and some of the 4th Corps District. In 1917 the division contained:

28th Reserve Division:
  56th Reserve Brigade:
    109th Reserve Regiment
    110th Reserve Regiment
    111th Reserve Regiment
  Cavalry:
    3/22nd Dragoon Regiment
  124th Artillery Command[25]:
    29th Reserve Field Artillery Regiment (9 btrys)

---

20 Joined in July 1916 from the 111th Infantry Division.

21 The 1st Battalion was the former 76th Ersatz Field Artillery Battalion, originally with the 55th Mixed Ersatz Brigade.

22 *Stein* says that the 9/28th Pioneer Battalion for some unspecified time, perhaps in 1915/16, also served with this division. It came from the 28th Infantry Division and the 18th Army and would subsequently go to the 15th Reserve Division and finally the 103rd Infantry Division.

23 Left for the 54th Reserve Division, according to *Stein*.

24 Raised in February 1916 from the 34th Heavy, the 144th Medium and the 246th Light Trench Mortar Sections.

328th Pioneer Battalion:
  1/, 4/16th Pioneer Battalion
  228th Trench Mortar Company
Signals:
  428th Telephone Detachment
Transport:
  725th Light Munitions Column
Medical:
  514th Medical Company
  56th Reserve Field Hospital
  57th Reserve Field Hospital
  ? Veterinary Hospital

On 16 February 1918 the division was withdrawn from the lines and underwent intensive training in offensive operations. On 20 March it was in the lines again. It led the attack on Fayet on 21 March and continued operations in the Battle of the Somme until 17 April. It was known to have suffered heavy losses – up to 75 % in many units – and received 2,500 men as reinforcements by the 17th. The division rested until 23 May and went into the lines, attacking on the Aisne Front, southwest of Craonne. It advanced 60 kilometers to the Marne in these attacks. On 3 to 7 June the division rested and went back to the lines near Chateau Thierry, facing the 2nd US Division. It suffered heavy losses at Belleau Wood and was withdrawn on 3 July. On 18 July it was engaged south of Soissons and remained in action until 1 August. It was rested until mid-August. It then went to the Champagne and moved between the Argonne and Meuse sectors as necessary until late September. It was engaged near Cunel from 3 to 18 October and was then withdrawn. The average company strength was about 40 to 50 men at that time. It was last known to be near Villiers devant Dun. In 1918 the division contained:

28th Reserve Division:
  56th Reserve Brigade:
    109th Reserve Regiment
    110th Reserve Regiment
    111th Reserve Regiment
  Cavalry:
    3/22nd Dragoon Regiment
  144th Artillery Command:
    29th Reserve Field Artillery Regiment (9 btrys)
    7/ & 9/2nd Bn/1st Guard Foot Artillery Regiment
    766th Light Ammunition Column
    918th Light Ammunition Column
    1366th Light Ammunition Column
328th Pioneer Battalion:
  4/16th Pioneer Battalion
  1st Ersatz Company, 16th Pioneer Battalion
  72nd Searchlight Section
428th Signals Command:
  428th Telephone Detachment
  160th Wireless Detachment
Medical:
  514th Medical Company
  56th Reserve Field Hospital
  57th Reserve Field Hospital
  428th Veterinary Hospital
Transport:
  725th Motor Transport Column

29th Infantry Division: The 29th Division, recruited in Baden and the Upper Alsace in the 14th Corps District, began the campaign with its three peacetime infantry brigades. Its first combat was west of Mülhausen/Mulhouse in German Alsace on 9 August holding up the French 1914 invasion. On 20 August it fought near Zabern/Savere. It then moved into Lorraine, crossed into France and suffered heavy casualties in the battle on the Meurthe. In October it was transferred to north of Arras and remained there until May 1915. From October 1914 to May 1915 the 29th Division suffered again heavy losses in fighting on the Notre Dame de Lorette Plateau. Casualties were tremendous: the 3/169th

---

[25] According to *Cron*, the 24th Reserve Field Artillery Brigade became the 124th Artillery Command on 16 February 1917. However, on 13 April the 144th Artillery Command joined but *Cron* does not relate what happened to the 124th.

Infantry Regiment had 38 men left and the 142nd Infantry Regiment had lost 44 officers and 2,603 men by the end of November 1914! In August 1914 the division contained:

29th Infantry Division:
- 57th Infantry Brigade:
  - 113th Infantry Regiment
  - 114th Infantry Regiment
- 58th Infantry Brigade:
  - 112th Infantry Regiment
  - 142nd Infantry Regiment
- 84th Infantry Brigade:
  - 169th Infantry Regiment
  - 170th Infantry Regiment

Cavalry:
- 1/, 2/, 3/, 5/22nd Dragoon Regiment[26]

29th Field Artillery Brigade:
- 30th Artillery Regiment
- 76th Artillery Regiment

Engineers:
- 1/14th Pioneer Battalion
- 29th Divisional Bridging Train
- 1st & 3rd Medical Companies

In March 1915 the 84th Infantry Brigade was sent to the new 52nd Division. Between 8 to 13 May the division was very heavily engaged and the 114th Infantry Regiment lost another 1,000 men. The rest of the division suffered similar losses. From January to March 1915 the division contained:

29th Infantry Division:
- 57th Infantry Brigade:
  - 113th Infantry Regiment
  - 114th Infantry Regiment
- 58th Infantry Brigade:
  - 112th Infantry Regiment
  - 142nd Infantry Regiment
- 84th Infantry Brigade[27]:
  - 169th Infantry Regiment
  - 170th Infantry Regiment

Cavalry:
- 1/, 2/, 3/, 5/22nd Dragoon Regiment

29th Field Artillery Brigade:
- 30th Artillery Regiment
- 76th Artillery Regiment

Engineers:
- 1/14th Pioneer Battalion
- 29th Divisional Bridging Train

Signals:
- 29th Telephone Detachment

In May 1915 the division was sent to the area of Lens, Pont à Vendin and Henin-Lietard to rest and recover. It returned to the lines on 25 May and suffered again heavy losses. On 13 June it was relieved and sent to the vicinity of Rheims where it remained until November. It sent one battalion of the 113th Infantry Regiment to fight in the Champagne battle. On 19/20 October the 112th and 142nd Infantry Regiments exploited a gas attack in the La Pompelle-Prosnes sector. It suffered insignificant losses. Until the end of 1915 the division remained in the Champagne near Tahure.

The division had no significant action during the winter of 1915/1916. In September 1916 it was transferred to the Somme. In this period it seems to have become a "flying division", dealing with emergencies along the front. Also in September the 114th Infantry Regiment left the division and was reassigned to the 212th Division and then later to the 199th Division. On 4 October the division went into action on the Somme, east of Cléry. It remained there until February 1917. In autumn 1916 each

---

26 Mobilized on 2 August 1914, leaving at home (Bruchsal, 14th Corps District) the 4th Sqn as its Ersatz Sqn . The 5th Sqn was designated 4th Field Sqn. The regiment left this division on 1 April 1915 and was distributed on 6 April as 1st and 2nd Half Regiments to the 115th and 111th Infantry Divisions. It is not known which cavalry units, if any, served with the division after this regiment left and before the 4/5th Jäger zu Pferd Regiment joined in January 1916.

27 Renumbered on 4 March 1915 as 104th Infantry Brigade of the 52nd Infantry Division. Joined on 10 April 1917 the 228th Infantry Division.

infantry regiment reformed themselves into four battalions in place of the establishment three. In 1916 the division contained:

29th Infantry Division:
- 57th Infantry Brigade[28]:
  - 113th Infantry Regiment
  - 114th Infantry Regiment
- 58th Infantry Brigade:
  - 112th Infantry Regiment
  - 142nd Infantry Regiment
- Cavalry:
  - 4/5th Jäger zu Pferd Regiment[29]
- 29th Field Artillery Brigade:
  - 30th Artillery Regiment
  - 76th Artillery Regiment
- Engineers:
  - 1/14th Pioneer Battalion
  - 29th Trench Mortar Company[30]
  - 29th Divisional Bridging Train
- Signals:
  - 29th Telephone Detachment
- Attached:
  - 29th Divisional Machine Gun Company
  - 60th Labor Battalion

On 31 January 1917 the division's infantry regiments were again reduced from four battalions each to three. The men from the disbanded fourth battalions were distributed to the other three battalions. On 15 February 1917 the division rested north of St. Quentin, but spent its time working on the construction of the Hindenburg Line. On 1 April it was sent to the Champagne and was engaged south of Nauroy on the 17th. It suffered heavy losses. In late April it received reinforcements of new drafts and men from the disbanded 626th Infantry Regiment. In mid-June the division was rested and again received large drafts of new men. Between 10 and 12 July the division moved to Verdun and entered the sector of Avoucourt Wood. It was attacked by the French on 17 July, suffering heavy losses. The division launched a major attack on 1 August to retake the positions lost on 17 July. The French engaged it on 20 August and in its counterattack it lost heavily again. In late October the division was withdrawn and reorganized. It had lost so heavily that the Class of 1918 was called forward with only two months' instruction. From 6 October to the last of December they remained in the lines along the right bank of the Meuse. In 1917 men serving in units of the 4th Corps District who were at home in the 14th were drafted into the division. In 1917 the division contained:

29th Infantry Division:
- 58th Infantry Brigade:
  - 112th Infantry Regiment
  - 113th Infantry Regiment
  - 142nd Infantry Regiment
- Cavalry:
  - 4/5th Jäger zu Pferd Regiment
- 130th Pioneer Battalion:
  - 1/, 5/14th Pioneer Battalion
  - 29th Trench Mortar Company
- 67th Artillery Command[31]:
  - 30th Artillery Regiment
- Signals:
  - 29th Telephone Detachment
- Medical:
  - 36th Medical Company
  - 263rd Field Hospital
  - 266th Field Hospital
  - 29th Veterinary Hospital
- Transport:
  - 562nd Motor Transport Column

---

[28] Joined the 212th Infantry Division on 15 September 1916 and was disbanded on 31 January 1917.

[29] The *Histories* attributes the entire regiment to this division in 1916. According to *Voigt* and *Stein*, however, say just the 4th Sqn served with this division.

[30] Raised in December 1915 from the 89th Heavy and the 201st Medium Trench Mortar Detachments and the 123rd Light Trench Mortar Section. Distributed in autumn 1918 among the 112th, 113th and 142nd Infantry Regiments.

The division remained in the Beaumont sector until about 1 April 1918. It rested near Montmedy from 6 to 25 April. The division was rebuilt and at this time its infantry companies averaged 180 to 200 men. In late April the division was sent to Belgium and moved into the lines northeast of Mont Kemmel during the night of 1/2 May. While at Kemmel the division lost heavily from Allied artillery fire. It was withdrawn about 15 May and rested until 14 June. In the night of 14/15 June it returned to the lines east of Langemarck. On 15 July the division was withdrawn until the 25th, when it went to the Laon district, and then on to the Vesle front.

From 1 to 31 August the division held the Courlaudon sector, then withdrew to the Aisne Region. It was withdrawn about 8 September and rested until the 15th, then reengaged north of the Aisne from 16 to 24 September. In early October the division was sent to Fresnoy le Grand and went into the lines on 6 October. On 7 to 8 October heavy French attacks drove the division back. After a short rest the division was reengaged on the 18th, south of Le Cateau. In November the division retreated in the direction of Avesnes, where it was last identified on 7 November. In 1918 the division contained:

29th Infantry Division:
- 58th Infantry Brigade:
  - 112th Infantry Regiment
  - 113th Infantry Regiment
  - 142nd Infantry Regiment
- Cavalry:
  - 4/5th Jäger zu Pferd Regiment
- 67th Artillery Command:
  - 30th Artillery Regiment
  - 2nd Bn/9th Reserve Foot Artillery Regiment
  - 734th Light Ammunition Column
  - 827th Light Ammunition Column
  - 932nd Light Ammunition Column

130th Pioneer Battalion:
- 1/, 5/14th Pioneer Battalion
- 29th Trench Mortar Company
- 185th Searchlight Section

29th Signal Command:
- 29th Telephone Detachment
- 31st Wireless Detachment

Medical and Transport:
- 36th Medical Company
- 263rd Field Hospital
- 266th Field Hospital
- 29th Veterinary Hospital
- 562nd Motor Transport Column

29th Landwehr Division: According to *Cron*, no 29th Landwehr Division was ever on the establishment – apparently another misinformation of the Allies. The 29th Landwehr Brigade remained independent throughout the war. However, it was reinforced from time to time.

The 29th Mixed Landwehr Brigade or Wynecken Brigade was formed as the war began and entered Belgium on 20 August 1914. In September it was at Berry au Bac. It contained in 1914:

29th Mixed Landwehr Brigade:
- 28th Landwehr Regiment[32]
- 29th Landwehr Regiment[33]

Cavalry:
- 1st Landwehr Squadron, 8th Corps District[34]

Artillery:
- Landsturm Battery, 8th Corps District

---

31 The 29th Field Artillery Brigade was converted into the 29th Artillery Command on 16 February 1917, but left at that date the division for the 108th Infantry Division. Instead the 67th Artillery Command joined 16 February 1917, which had been raised as 67th Field Artillery Brigade on 4 October 1917 without *Cron* mentioning to which division that had been attached in between.

32 Raised in August 1914 in the Rhenish 8th Corps District by the 25th Infantry Regiment (Staff, 2nd Bn; at Aachen/Aix la Chapelle; 1st Bn; at Cologne) and the Neuss District Command (3rd Bn).

33 Raised in August 1914 in the Rhenish 8th Corps District at Cologne.

From January to March 1915 the brigade comprised:

29th Mixed Landwehr Brigade:
28th Landwehr Regiment
29th Landwehr Regiment
Cavalry:
Ersatz Cavalry Regiment, 29th Landwehr Brigade[35]

Artillery:
247th Artillery Regiment[36]
Attached:
11th Landsturm Battalion

During early 1915 also the future 89th Cavalry Regiment[37] might have been attached to the brigade. The 29th Landwehr Brigade was transferred to the Eastern Front at the end of March 1915. In April and May it was in action in Poland, south of Rawa. The brigade was heavily reinforced in 1915 by troops seconded from other sources, so that it was referred to from 1 June to 20 October 1915 as Wyneken[38] Truppenabteilung (Detachment of Troops) or Wyneken Division. When the 1915 offensive began, the brigade sent the 28th Landwehr Regiment to Courland. It fought at Kaluzem, then Vilna, and finally rejoined the brigade near Kalnzem, north of Mitau, in early November. The 29th Landwehr Regiment fought at first at Vilna and then east of Olai in October. In November it rejoined the 28th Landwehr Regiment.

The 29th Mixed Landwehr Brigade continued operations in Courland in 1916 and would remain there until March 1917. In October it was joined by the 427th Infantry

---

34 Raised in August 1914 by the 7th Hussar Regiment. Joined later the 2nd, 11th and Bug Line-of Communication Inspections and became on 9 October 1916 the 4/3rd Heavy Reserve Cavalry (Schützen) Regiment.

35 Raised in August 1914 for the Königsberg Fortress Main Reserve with the Ersatz Sqns of the 8th and 12th Uhlans and the 10th Jäger zu Pferd – the 2nd to 4th Mobile Ersatz Sqns, 1st Corps District. Also called Mobile Ersatz Cavalry Regiment, 1st Army Corps. Served with the improvised Charisius Cavalry Brigade, the Königsberg Ersatz Brigade, the Wynecken Brigade, the 75th and 3rd Reserve Divisions and the 3rd Cavalry Brigade. Became, on 7 August 1915, the 89th Cavalry Regiment and served apparently with the 109th Infantry Division, again with the 29th Infantry Brigade and then with the 89th Infantry Division.

36 Including the former Landsturm Field Artillery Battery, 8th Corps District.

37 Improvised in early August 1914 as Mobile Ersatz Cavalry Regiment, 1st Corps District, with the following Ersatz Sqns: 2/8th (also called 2nd Mobile Ersatz Sqn, 1st Corps District) and 1/12th Uhlans (also called 3rd Mobile Ersatz Sqn, 1st Corps District) and 4/10th Jäger zu Pferd (also called 4th Mobile Ersatz Sqn, 1st Corps District). Those would be replaced by new Ersatz squadrons at home. Later, also the 6th Mobile Ersatz Sqn, 1st Corps District joined (raised in early April 1915 as Combined Königsberg Ersatz Sqn or Pein Sqn by drafts on the Ersatz Sqns of the 3rd Cuirassiers, 1st Dragoons, 8th and 12th Uhlans, 9th and 10th Jäger zu Pferd Regiments).

There are contradictions in *Stein's* papers concerning the whereabouts of this regiment. According to one paper, the regiment originally belonged to the Königsberg Main Reserve, then to the improvised Charisius Cavalry Brigade and the Königsberg Ersatz Brigade, the 38th Reserve Corps, the 75th and 3rd Reserve Divisions and the 3rd Cavalry Brigade before being numbered. According to another it started with the Königsberg War Garrison, went on to the Wynecken Landwehr Brigade, the 109th Infantry Division, the 29th Landwehr Brigade – at that time called Mobile Ersatz Cavalry Regiment, 29th Landwehr Brigade – and finally to the 89th Infantry Division. Numbered as 89th Cavalry Regiment and put onto the establishment on 7 August 1915.

Dismounted in autumn 1916 and became the 89th Cavalry (Schützen) Regiment, the Machine Gun Sqn/20th Dragoon Regiment and the 2nd Guard Machine Gun Detachment joining as 1st and 2nd Machine Gun Sqns. Regiment disbanded on 29 September 1918 and distributed among the 5th Cuirassier and the 4th Reserve Uhlan Schützen Regiment.

38 Lieutenant General Wyneken was the commanding officer from 19 March 1915 to 3 November 1917.

Regiment, but that was returned to the 9th Landwehr Brigade in January 1917. In January 1916 the brigade contained:

29th Mixed Landwehr Brigade:
 28th Landwehr Regiment
 29th Landwehr Regiment
 Cavalry:
 4/8th Uhlan Regiment[39]
 Artillery:
 247th Artillery Regiment
 Engineers:
 6th Landsturm Pioneer Company, 3rd Corps Dist[40]
 182nd Trench Mortar Company[41]
 Signals:
 289th Wireless Detachment
 302nd Wireless Detachment
 Medical:
 One Medical Company
 One Field Hospital
 One Veterinary Hospital

Organization of the 29th Landwehr Brigade in 1917 is not known. It was on the Russian Front. In mid-March 1917 the 29th Landwehr Brigade was sent to Kreuzburg west of Jakobstadt in Courland. It appears to have served as a reserve in the operations against Riga. In November 1917 the brigade was used, according to the *Histories*, to form the 29th Landwehr Division, but despite this, men were drafted from the division to provide replacements for the 77th Reserve Division. According to *Cron*, however, from 19 December 1917 onwards the brigade's establishment provided for two Landwehr infantry regiments, one cavalry squadron, one field artillery regiment, and one company, each, of pioneers and trench mortars.

In 1918 the 29th Landwehr Division (or rather 29th Landwehr Infantry Brigade) occupied a very extended position north of Kreuzburg. It remained there until March when it was sent to the Polotsk-Vitebsk Region, and then on to Estonia. It was in Estonia in the beginning of May and probably remained there until the end of the war serving as the Dünaburg Commandants Office. In 1918 the division contained:

29th Landwehr Division:
 29th Landwehr Infantry Brigade:
 28th Landwehr Regiment
 29th Landwehr Regiment
 Cavalry:
 4/8th Uhlan Regiment
 Artillery:
 247th Artillery Regiment
 Engineers:
 6th Landsturm Pioneer Company, 3rd Corps District[42]
 One Trench Mortar Company
 Medical:
 555th Medical Company
 327th Field Hospital
 10th Veterinary Hospital

30th Infantry Division: When the war began the 30th Division, in peace at home in the 15th Corps District at Strassburg/Strasbourg, protected the Upper Alsace from 8 to 13 August. It subsequently moved to south of Saarburg/Sarrebourg and there engaged the French on 20 August. In early September it was concentrated near Avricourt and then moved on to Hurtebise. It fought there from mid-September to mid-October 1914. In

39 Might only have joined in autumn 1916 from the 105th Infantry Division. Was seconded to the Commandant's Office of the Reval Fortress.

40 Joined from the 17th Cavalry Brigade and remained until the war's end.

41 Mobilized in November 1916 from the 125th, 126th and the 127th Trench Mortar Reinforcements. Joined at some time, which *Stein* does not specify, the 202nd Infantry Division.

42 Joined from the 17th Cavalry Brigade and remained until the war's end.

November the division was sent to Flanders and went into action south of Ypres. It stayed there until January 1916. In August 1914 the division contained:

30th Infantry Division:
- 60th Infantry Brigade:
  - 99th Infantry Regiment
  - 143rd Infantry Regiment
- 85th Infantry Brigade:
  - 105th Saxon Infantry Regiment[43]
  - 136th Infantry Regiment
- Cavalry:
  - 2/ to 5/3rd Jäger zu Pferd Regiment[44]

30th Field Artillery Brigade:
- 51st Field Artillery Regiment
- 84th Field Artillery Regiment

Support Troops:
- 1/15th Pioneer Battalion
- 30th Divisional Bridging Train
- 1st & 2nd Medical Company

Through 1915 the division remained in the vicinity of Ypres. In April the 136th Infantry Regiment was detached and reassigned to the new 115th Division and the 61st Brigade was disbanded. From January to April 1915 the division contained:

30th Infantry Division:
- 60th Infantry Brigade:
  - 99th Infantry Regiment
  - 143rd Infantry Regiment
- 85th Infantry Brigade[45]:
  - 105th Infantry Regiment
  - 136th Infantry Regiment
- Cavalry:
  - unknown

30th Field Artillery Brigade:
- 51st Field Artillery Regiment
- 84th Field Artillery Regiment

Engineers:
- 1/15th Pioneer Battalion
- 30th Divisional Bridging Train

Signals:
- 30th Telephone Detachment

In January 1916 elements of the division were sent to Verdun for the coming February offensive. Eventually the entire division was assembled and went into action on the Maucourt-Warcq front. The division suffered only limited losses, but the 105th Infantry, which was fighting elsewhere, with the 3rd Army Corps, suffered very heavy losses during this period. On 11 July the 99th and 143rd Infantry took part in a new offensive. On 8 August the 143rd attacked fortifications at Thiaumont and was decimated. A very slow process of replacements began at this point. At the end of September the division was sent to rest near Cambrai. After only a few days the division went into the lines at Sailly Saillisel and during the next month again suffered heavy losses. The 143rd Regiment lost about half its effectives in this period. In late November the division was withdrawn and returned to the Verdun Front. In early December it was sent to the Meuse, but did not take part in any significant combat for the rest of the year. In 1916 the division contained:

---

43 A regiment of the Saxon Army seconded to the peacetime Garrison of Strassburg/Strasbourg.

44 Mobilized in August 1914 with staff and 2nd to 5th Sqns. The 1st Sqn was left behind as Ersatz Sqn and part of the Neubreisach/Neuf Brisach Fortress Garrison stationed at Eichstetten, 15th Corps District. Already at the end of September 1914 the regiment joined the 4th Cavalry Division and the new 39th Cavalry Brigade and fought on the Eastern Front from 11 November 1914 onwards. In 1915 the regiment formed a machine gun troop, which was expanded in 1916 into a machine gun squadron. It later joined the 4th Reserve Uhlan (Schützen) Regiment. After leaving the brigade in autumn 1916 the regiment became part of the army troops. As such it served in the 12th Line-of-Communications Inspection and the Lithuanian Military District fighting brigands until the war's end.

45 Disbanded on 3 April 1915.

30th Infantry Division:
- 60th Infantry Brigade:
  - 99th Infantry Regiment
  - 105th Infantry Regiment
  - 143rd Infantry Regiment
- Cavalry:
  - 2/8th Reserve Hussar Regiment[46]
- 30th Field Artillery Brigade:
  - 51st Field Artillery Regiment
  - 84th Field Artillery Regiment
- Engineers:
  - 1/15th Pioneer Battalion
  - 30th Trench Mortar Company[47]
  - 30th Divisional Bridging Train
- Signals:
  - 30th Telephone Detachment
- Attached:
  - 30th Anti-Aircraft Detachment

On 1 March 1917 the division was sent to the Champagne and suffered from the Allied spring offensive in that Region in early May. It would remain in the lines there until August. The division suffered, during this period, extremely heavy losses, with the 105th Infantry being reduced to 400 men. That regiment was rebuilt in the Argonnes. In late August the division was sent to the Meuse near Forges-Bethincourt. On 24 October it was again relieved and when the British attacked Cambrai on 20 November it was hastily recalled. It went into the line on 23 November and remained there until 10 December. At this time the division was sent to the Sedan area in the Champagne, and rested there until mid-January 1918. While the 105th Infantry Regiment in 1917 still received its replacements from Saxony, the other two Prussian infantry regiments were mostly recruited of Westphalians from the 7th Corps District. In 1917 the division contained:

30th Infantry Division:
- 60th Infantry Brigade:
  - 99th Infantry Regiment
  - 105th Infantry Regiment
  - 143rd Infantry Regiment
- Cavalry:
  - 2/8th Reserve Hussar Regiment
- 30th Artillery Command:
  - 84th Field Artillery Regiment
- 131st Pioneer Battalion:
  - 1/, 5/15th Pioneer Battalion
  - 30th Trench Mortar Company
- Signals:
  - 30th Telephone Detachment
- Medical:
  - 37th Medical Company
  - 39th Medical Company
  - ? Field Hospital
  - ? Veterinary Hospital
- Transport:
  - ? Light Munitions Column
- Attached:
  - 30th Anti-Aircraft Detachment

Until 27 March 1918 the division remained in the Champagne. At that time it was transferred to the Montdidier area, reaching Framicourt on 12 April. On 16 May it was withdrawn having suffered 30% casualties in the meantime. The division was then rebuilt and trained south of Roy. On 12 June it went into the front near Courcelles and again suffered heavy losses. It was withdrawn on 22 June and sent to the Champagne. There it participated in the Champagne offensive until 10 August, when it was sent to rest near Lens. On 23 September the division moved to the Cambrai Region and remained there until 28 October. It then rested near Maubeuge. In 1918 the division contained:

---

46 According to *Stein* when this regiment was disbanded, on 17 October 1916, its 2nd Sqn joined this division. In 1918 it was probably stationed on the Maubert-Fontaine Training Ground and was disbanded itself on 21 August 1918.

47 Raised in November 1915 from the 43rd Heavy Trench Mortar Detachment, the 110th Medium and one third of the 208th Light Trench Mortar Sections and the 33rd Light Trench Mortar Supplement Detachment. Distributed in autumn 1918 among the division's infantry regiments.

30th Infantry Division:
60th Infantry Brigade:
99th Infantry Regiment
105th Infantry Regiment
143rd Infantry Regiment
Cavalry:
5/7th Dragoon Regiment[48]
30th Artillery Command:
84th Field Artillery Regiment
10th Bavarian Foot Artillery Battalion
885th Light Ammunitions Column
1369th Light Ammunitions Column
1394th Light Ammunitions Column
15th Pioneer Battalion:
1/, 5/15th Pioneer Battalion
30th Searchlight Section
30th Signal Command:
30th Telephone Detachment
161st Wireless Detachment
Medical:
30th Medical Company
275th Field Hospital
276th Field Hospital
30th Veterinary Hospital

31st Infantry Division: This division, recruiting in the 21st Corps District in the Sarre area of the Rhine Province and the Lower Alsace, had, in peacetime, one infantry regiment more than a normal infantry division. In August 1914 it was on the Lorraine frontier. It advanced to Dieuze where it was engaged on 20 August and the 174th Infantry Regiment lost half its effectives. The division advanced as far as Moyen-Domptail, before it then withdrew to Dieuze. About 18 September it was transferred to the Somme. It attacked towards Bouchoir on 6 October and in late October occupied the front at Fouquescourt-Chaulnes, which it then held until the end of January 1915. By that time the 174th Infantry Regiment had lost a total of 81 officers and 3,521 other ranks from August 1914 onwards. In August 1914 the division contained:

31st Infantry Division:
32nd Infantry Brigade:
70th Infantry Regiment
174th Infantry Regiment
62nd Infantry Brigade:
60th Infantry Regiment
137th Infantry Regiment
166th Infantry Regiment
Cavalry:
1/, 2/, 4/, 5/7th Dragoon Regiment[49]
31st Field Artillery Brigade:
31st Field Artillery Regiment
67th Field Artillery Regiment
Support Troops:
1/27th Pioneer Battalion
31st Divisional Bridging Train
1st Medical Company

---

48 Joined apparently at some time in 1918 from the 10th Landwehr Division.

49 The *Histories*, attributes, in 1914, the 7th Uhlans to both this division as well as to the 42nd, which obviously is an error. *Voigt* also attributes the 7th Uhlan Regiment and the 7th Dragoons to the 42nd Infantry Division.

Regiment mobilized in August 1914 leaving behind the 3rd Sqn as Ersatz Sqn at Saarbrücken/Sarrbruck, 21st Corps District (itself mobile from 2 August 1914 to 31 January 1915). In January 1916 a machine gun troop was raised which later joined the 131st Infantry Regiment. In October 1916 a new machine gun troop was raised and attached to the 1st Sqn.

By November 1914, according to *Stein*, the 4th and 5th (= 3rd Field Sqn) formed together with the 3/ and 4/7th Uhlan Regiment (from 42nd Division) the improvised Flotow Cavalry Regiment, 4th Cavalry Division. In May 1915 the Flotow Cavalry Regiment was reduced to the Flotow Cavalry Detachment with the 4/7th Dragoons and 4/7th Uhlans, was renamed Dieringshofen Cavalry Regiment in July 1915 and disbanded at an unknown time.

The 2nd Sqn in December 1914 joined the improvised 1st Regiment/Keller Brigade (*Stein* does not report when this improvised brigade and its two regiment were dissolved again); Brigade Staff: Staff/7th Uhlan Regiment; 1st Regiment: 1/, 3/8th Dragoons (from the 5th Cavalry Division) and 2/7th Dragoons (from the 31st Infantry Division); 2nd Regiment: 1/, 2/, later also 3/4th Bavarian Light Cavalry Regiment (from the 2nd Bavarian Infantry Division), 5/7th Uhlan Regiment (from the 31st Infantry Division), 4/6th Uhlan Regiment (4th Field Sqn = 5th Sqn; from the 21st Infantry Division) and 3/6th Dragoon Regiment (from the 25th Infantry Division).

On 25 January 1915 the division was sent to the Eastern Front, leaving the 60th Infantry Regiment in France. In early February the division concentrated on the East Prussian frontier and became part of Hindenburg's Army. By 20 February it advanced as far east as Seporkin and took up position north of Grodno. It suffered heavy losses on 9 March in a Russian counterattack. From 29 March to 24 April the division participated in the fighting around Kalwariia and Mariampol. In the combats it again suffered heavily. The 1st Bn/166th Infantry Regiment lost 17 officers and 1,022 men and the 1st Company alone 5 officers and 336 men. Subsequently, the division was withdrawn and reorganized in late April. By the end of July the division was back in the lines around Mariampol.

The 137th Regiment was transferred to the 108th Division in the summer of 1915. In August 1915 the division participated in the offensive against Vilna and reached the vicinity of Smorgoni-Soly in October. The division was relieved for a short period in mid-October and by the 24th was back in the lines at Poslawy-Lake Narotch. It remained there until December 1917, when it returned to the Western Front. In January 1915 the division contained:

31st Infantry Division:
- 32nd Infantry Brigade:
  - 70th Infantry Regiment
  - 174th Infantry Regiment
- 62nd Infantry Brigade[50]:
  - 60th Infantry Regiment
  - 137th Infantry Regiment
  - 166th Infantry Regiment
- Cavalry:
  - 7th Dragoon Regt (unknown portion probably present)
- 31st Field Artillery Brigade:
  - 31st Field Artillery Regiment
  - 67th Field Artillery Regiment
- Engineers:
  - 1/27th Pioneer Battalion
  - 31st Divisional Bridging Train
- Signals:
  - 31st Telephone Detachment

During 1916 the division continued in its Lake Narotch sector. At the end of March it resisted to a Russian offensive and had very severe casualties. In 1916 the division contained:

31st Infantry Division:
- 32nd Infantry Brigade:
  - 70th Infantry Regiment
  - 166th Infantry Regiment
  - 174th Infantry Regiment
- Cavalry:
  - 1/, 5/7th Uhlan Regiment[51]
- 31st Field Artillery Brigade:
  - 31st Field Artillery Regiment
  - 67th Field Artillery Regiment
- Engineers:
  - 1/, 2/27th Pioneer Battalion
  - 31st Trench Mortar Company[52]
  - 31st Divisional Bridging Train
- Signals:
  - 31st Telephone Detachment
- Attached:
  - Field Recruit Depot, 31st Division
  - 153rd Cyclist Company

So, in 1914 and 1915 only the 1/7th Dragoon Regiment might have stayed with the 31st Infantry Division if the 5th Sqn (3rd Field Sqn) did not rejoin. Perhaps in summer 1916 the 1st and 2nd Sqns joined the 42nd Infantry Division the 2nd passing on to the 17th Landwehr Division. The 4th Sqn consecutively joined the 81st Reserve, the 44th Reserve and then the 201st Infantry Division where it srayed until the end of the war and the 5th Sqn consecutively passed to the 11th Bavarian Infantry Division, the 10th Landwehr Division where it remained for most of the war and finally to the 30th Infantry Division.

50 On 23 November 1916 to 218th Infantry Division.

51 The 1st might have probably joined in 1916 from the 42nd Infantry Division, passing on to the independent 9th Landwehr Brigade and then in 1917 to the 94th Infantry Division. The 5th Sqn

On 5 December 1917 the division was relieved from the Russian Front and entrained for Belgium on 16 December. It arrived on 21 December near Ghent and in January 1918 took up positions south of the Ypres-Roulers line. As the division's home area on the Sarre around Saarbruecken/Sarrebruck, St. Wendel and Hagenau/Haguenau in the 21st Corps District could not furnish enough recruits, by 1917 replacements were mostly Westphalians from 7th Corps District. In 1917 the division contained:

31st Infantry Division:
- 32nd Infantry Brigade:
  - 70th Infantry Regiment
  - 166th Infantry Regiment
  - 174th Infantry Regiment
- Cavalry:
  - 5/7th Uhlan Regiment
- 31st Artillery Command:
  - 31st Field Artillery Regiment
- 136th Pioneer Battalion:
  - 1/27th Pioneer Battalion
  - 31st Trench Mortar Company
  - 505th Searchlight Section
- Signals:
  - 31st Telephone Detachment
- Medical:
  - 262nd Medical Company
  - ? Field Hospital
  - ? Veterinary Hospital
- Transport:
  - ? Light Munitions Column

The division served in the Ypres Region until 4 April 1918 interrupted by short rests. It then moved to the Messines Front and was engaged from 10 to 12 April in the Battle of the Lys in the Bois Ploegsteert. It rotated into and out of the lines, frequently engaging in heavy action and suffered accordingly. The 3rd Bn/174th Regiment lost 60 to 70% of its effectives. Around 1 May the division left the Kemmel area and rested near Tourcoing until 6 May. Having returned to the Kemmel area it suffered heavy losses in a French attack on 21 May. It was relieved on the 24th and rested until 15 June, when it returned to a sector south of Ypres where it remained until 27 July.

In late July the division was sent to the Woevre Region. It took over the St. Mihiel sector on 29 July and remained there until 3 September. During the Battle of St. Mihiel, when the Americans attacked, the division was resting north of Dampvitoux. It returned to the lines near Thiaucourt and remained there from 14 to 28 October. It was then engaged near Imocourt and remained there until 11 November. In 1918 the division contained:

31st Infantry Division:
- 32nd Infantry Brigade:
  - 70th Infantry Regiment
  - 166th Infantry Regiment
  - 174th Infantry Regiment
- Cavalry:
  - 5/7th Uhlan Regiment[53]
- 31st Artillery Command:
  - 31st Field Artillery Regiment
  - 44th Foot Artillery Battalion
  - 1005th Light Ammunitions Column
  - 1011th Light Ammunitions Column
  - 1012th Light Ammunitions Column
- 93rd Pioneer Battalion:
  - 1/27th Pioneer Battalion
  - 3/32nd Reserve Pioneer Battalion
  - 31st Trench Mortar Company
  - 40th Searchlight Section
- 31st Signal Command:
  - 31st Telephone Detachment
  - 116th Wireless Detachment

---

could have joined even in February 1915 from the improvised Keller Cavalry Brigade. It stayed until the end of 1917 and served on the lines-of-communications during 1918 according to *Voigt*.

52 Raised in February 1916 from the 24th Heavy and the 165th Medium Trench Mortar Detachments. Joined in autumn 1918 the 174th Infantry Regiment.

53 It stayed until the end of 1917 and served on the lines-of-communications during 1918 according to *Voigt*.

Medical:
- 262nd Medical Company
- 326th Field Hospital
- 333rd Field Hospital
- 161st Veterinary Hospital

Transport:
- 564th Motor Transport Column

Attached (Elements attached 15 May 1918):
- 24th Machine Gun Sharpshooter Battalion
- 2nd/, 3rd Bn/44th Field Artillery Regiment
- 121st Foot Artillery Regiment
- 1st Bn/44th Foot Artillery Regiment
- 2nd Bn/6th Foot Artillery Regiment
- 1st Bn/74th Foot Artillery Regiment
- 134th Artillery Observation Section
- 75th Sound Ranging Detachment
- 134th Balloon Squadron
- 105th Balloon Squadron
- 256th Reconnaissance Flight

32nd Infantry Division: The 32nd Infantry Division belonged to the Saxon Army and the 12th Army Corps. Even on 2 August 1914 its 64th Brigade moved to the Luxembourg frontier to cover the German concentration on that frontier. The division entered Belgium on 13 August and France on 23 August. On 28 August it was engaged at Signy l'Abbaye and then moved on Châlons. It fought in the Battle of the Marne. Subsequently the division attacked in the vicinity of Rheims in the Champagne. When the front stabilized it was near Berry au Bac and remained there until July 1916. In August 1914 the division contained:

32nd Infantry Division:
- 63rd Infantry Brigade:
  - 1st/, 2nd/, 3rd Bn/102nd Infantry Regiment
  - 1st/, 2nd/, 3rd Bn/103rd Infantry Regiment
  - 12th Jäger Battalion
- 64th Infantry Brigade:
  - 1st/, 2nd/, 3rd Bn/177th Infantry Regiment
  - 1st/, 2nd/, 3rd Bn/178th Infantry Regiment
- Cavalry:
  - 1/, 2/, 4/, 5/18th Hussar Regiment[54]
- 32nd Field Artillery Brigade:
  - 1st Bn/28th Field Artillery Regiment (3 btrys, 6-77mm guns ea)
  - 2nd Bn/28th Field Artillery Regiment (3 btrys, 6-77mm guns ea)
  - 1st Bn/64th Field Artillery Regiment (3 btrys, 6-77mm guns ea)

---

[54] Mobilized in August 1914 leaving the 3rd Sqn behind at Grossenhain, Saxony, as Ersatz Sqn. Apparently, the 1st Sqn already in December 1914 was detached (it might have joined the Rüxleben Cavalry Regiment as did the 5/20th Hussars) and joined in May 1915 together with the 5/20th Hussars the improvised Cavalry Detachment, 123rd Infantry Division. *Stein* does not report how long this existed. *Voigt* says, though, that this lasted until October 1915. The squadron later, perhaps still in 1915, joined (according to *Stein*, not mentioned by *Voigt*) the 187th and in 1917 (according to *Voigt*) the 212th Infantry Division where it stayed for a longer time. However, at some time not mentioned by *Stein*, either, it left for the 9th Army, afterwards served with the 303rd Divisional Command for Special Purposes and then with the Mackensen Army Group.

The remainder of the regiment left its division and was transferred to the east in October 1915 where it served on the lines-of-communication until June 1916. At some time before September 1916 it joined the 3rd Cavalry Division (according to *Voigt* the 9th instead), went on 4 September 1916 to Rumania and was transferred, when the 3rd Cavalry Division was disbanded on 2 November 1916, to the Siebenbürgen (= Transylvanian) Cavalry Brigade. In 1917 the regiment was split up. The 2nd Sqn joined the 241st and the 4th the 96th Infantry Division where they stayed until the end of the war. The 5th, though, consecutively joined the 9th Army, the 303rd Divisional Command for Special Purposes and finally the 212th Infantry Division. It was disbanded in 1918.

Probably in 1916 a machine gun squadron had been added. When the regiment was distributed in 1917 it joined the 9th Army and was seconded at first to the Turkish 15th Infantry Division, in September1917 to the Austro Hungarian 145th Infantry Brigade and in May 1918 to the 212th Infantry Division.

2nd Bn/64th Field Artillery Regiment
(3 btrys, 6-105mm howitzers ea)
Four Light Munitions Columns

Support Troops:
2/, 3/12th Pioneer Battalion[55]
32nd Divisional Bridging Train
2nd Medical Company

In April 1915 the 178th Infantry Regiment was detached to the newly forming 123rd Division. In 1915 the division contained:

32nd Infantry Division:
63rd Infantry Brigade:
102nd Infantry Regiment
103rd Infantry Regiment
64th Infantry Brigade[56]:
177th Infantry Regiment
178th Infantry Regiment
Cavalry:
18th Hussar Regiment[57]

32nd Field Artillery Brigade:
28th Field Artillery Regiment
64th Field Artillery Regiment
Engineers:
2/12th Pioneer Battalion
32nd Divisional Bridging Train
Signals:
32nd Telephone Detachment

The division remained in its quiet sector until the end of June 1916. At that time the division was apparently sent to the Somme as reserves facing the Franco-British offensive. At the beginning of July one battalion went into action. In late July the 102nd and 103rd Infantry Regiments were sent to the provisional Franke Division, which fought on the Somme until 10 September. These regiments suffered heavy losses during this period. On 4 September the 177th Infantry Regiment was sent to the Franke Division and between 4 and 10 September it lost 1,600 men. On 10 September the Franke Division was withdrawn and dissolved.

The 32nd Infantry Division was reformed with the 102nd, 103rd and 177th Infantry Regiments, reorganized, and sent to the Argonnes. In early November it returned to the Somme and occupied a sector northeast of Clery until March 1917. In 1916 the division contained:

32nd Infantry Division:
63rd Infantry Brigade:
102nd Infantry Regiment
103rd Infantry Regiment
Cavalry:
4/20th Hussar Regiment[58]
32nd Field Artillery Brigade:
28th Field Artillery Regiment
64th Field Artillery Regiment

Engineers:
2/12th Pioneer Battalion
32nd Trench Mortar Company[59]
32nd Divisional Bridging Train
Signals:
32nd Telephone Detachment
Attached:
115th Labor Battalion

---

55 3/12th Pioneer Battalion joined the 23rd Infantry Division.

56 According to *Cron*, on 1 February 1915, it was renamed the 245th Infantry Brigade. In March 1915 to 123rd Infantry Division.

57 Left in October 1915.

58 The *Histories* attributes three squadrons of that regiment, but according to *Voigt* and *Stein*, only the 4th served in this division from July 1916 until the end of the war joining from the 23rd Infantry Division.

59 Raised in January 1916 from one third of the 143rd Medium Trench Mortar Detachment and the 194th Medium Trench Mortar Section.

The division left the Somme in late March 1917, moving to the Champagne. It was engaged in the Mont sans Nom sector on 17 and 18 April and suffered heavily. The division was withdrawn from the lines until 5 May when it went into the lines west of Tahure. About 10 June it was sent to Flanders and took up positions on the Ypres Front. It remained there until January 1918 interrupted by spells of rest. In 1917 the division contained:

32nd Infantry Division:
- 63rd Infantry Brigade:
  - 102nd Infantry Regiment
  - 103rd Infantry Regiment
  - 177th Infantry Regiment
- Cavalry:
  - 4/20th Hussar Regiment
- 32nd Artillery Command:
  - 64th Field Artillery Regiment
- 140th Pioneer Battalion:
  - 2/. 5/12th Pioneer Battalion
  - 32nd Trench Mortar Company
- Signals:
  - 32nd Telephone Detachment
- Medical:
  - 28th Medical Company
  - 308th Field Hospital
  - ? Veterinary Hospital
- Transport:
  - ? Light Munitions Column
- Attached:
  - 61st Machine Gun Sharpshooter Battalion

On 15 January the division was relived and rested near Tournai until early March. It then went into the lines northwest of Lille about 1 March. On 4 April it moved south and was engaged at Fleurbaix on 9 April. It remained in the lines until 8 May and suffered heavy losses during that period. It received a reinforcement of 450 men on 17 April. From 8 to 26 May it rested behind the Lys Front. It was posted west of Merville until the end of June. On 4 July it arrived in Lorraine and received a draft of 1,500 men. From 15 July to 1 October the division occupied the quiet Eix-Bezonveaux sector. From 9 to 24 October it was engaged on the right bank of the Meuse in the Bois de Morey, suffering heavy losses and having some companies reduced to as little as 15 men. Subsequently the division returned to the Bezonveaux sector and stayed there until the end of the war. In 1918 the division contained:

32nd Infantry Division:
- 63rd Infantry Brigade:
  - 102nd Infantry Regiment
  - 103rd Infantry Regiment
  - 177th Infantry Regiment
- 32nd Artillery Command:
  - 64th Field Artillery Regiment
  - 80th Foot Artillery Battalion
  - 942nd Light Ammunitions Column
  - 1063rd Light Ammunitions Column
  - 1266th Light Ammunitions Column
- 140th Pioneer Battalion:
  - 2/. 5/12th Pioneer Battalion
  - 32nd Trench Mortar Company
  - 220th Searchlight Section
  - 3rd Reserve Company, 12th Pioneer Battalion[60]
- Cavalry:
  - 4/20th Hussar Regiment
- 32nd Signal Command:
  - 32nd Telephone Detachment
  - 20th Wireless Detachment
- Medical:
  - 28th Medical Company
  - 116th Field Hospital
  - 308th Field Hospital
  - 32nd Veterinary Hospital
- Transport
  - 565th Motor Transport Column

33rd Infantry Division: The 33rd Infantry Division, at home in the Metz Fortress in Lorraine, in the 16th Corps District, began mobilizing already on 29 July. It went into France via Audun le Roman and to the north of Verdun, crossing the Meuse on 1

[60] Joined the 212th Pioneer Battalion towards the end of the war.

September and advancing as far as Heippes, 20 kilometers south of Verdun. After the Battle of the Marne the division took up positions in the Argonnes. The division had lost substantial casualties. By 24 September the 98th Infantry Regiment was reduced to 13 officers and 982 men. A three-battalion regiment had an establishment strength of 85 officers and 3,304 other ranks! In August 1914 the division contained:

33rd Infantry Division:
- 66th Infantry Brigade:
  - 98th Infantry Regiment
  - 130th Infantry Regiment
- 67th Infantry Brigade:
  - 135th Infantry Regiment
  - 144th Infantry Regiment
- Cavalry:
  - 2/ to 5/12th Jäger zu Pferd Regiment[61]

33rd Field Artillery Brigade:
- 33rd Field Artillery Regiment
- 34th Field Artillery Regiment

Support Troops:
- 1/16th Pioneer Battalion
- 33rd Divisional Bridging Train
- 1st & 3rd Medical Company

The division remained in the Argonnes from September 1914 throughout 1915, not leaving until mid-August 1916. In 1915 the division contained:

33rd Infantry Division:
- 66th Infantry Brigade:
  - 98th Infantry Regiment
  - 130th Infantry Regiment
- 67th Infantry Brigade:
  - 135th Infantry Regiment
  - 144th Infantry Regiment
- Cavalry:
  - 4/, 5/12th Jäger zu Pferd Regiment

33rd Field Artillery Brigade:
- 33rd Field Artillery Regiment
- 34th Field Artillery Regiment

Engineers:
- 1/16th Pioneer Battalion
- 33rd Divisional Bridging Train

Signals:
- 33rd Telephone Detachment

Around 10 August 1916 the division left the Argonnes and moved to Verdun, east of Fleury. While in this sector the division was in almost constant combat and suffered heavy losses. The division remained in the lines at Verdun until September. When it withdrew the 144th Infantry Regiment was sent to the newly forming 223rd Division. From September to December the division was in the Argonnes to be rebuilt and reorganized. At the end of September 1916 the 40th Landwehr Regiment was detached from the 12th Landwehr Division and sent to the 33rd Infantry Division. At the end of October the 4th Battalions of the division's infantry regiments were broken up and distributed throughout the remaining battalions, as well as the 4th Bn/27th Landwehr Regiment. About 15 December the division was sent to the Somme where it remained

---

61 Mobilized in August 1914 leaving the 1st Sqn at home, in Hanover, 10th Corps District, as the Ersatz Sqn. The 2nd and 5th Sqns attacked the French in a mounted charge on 22 August 1914 with negligible losses. However, the regiment served until mid-September 1914 in the trenches as infantry, then until the end of the year as military police on the lines-of-communications and until March 1915 again as infantry. In March/April 1915 it was split into two half regiments. According to *Stein*, the staff, 4th and 5th Sqns stayed with the 33rd Infantry Division, while the 2nd and 3rd Sqns joined the 34th Infantry Division at first and soon passed to the 121st Infantry Division. When the 2nd and 3rd Sqns left the 34th Infantry Division they were replaced by the 5th Sqn while the 4th stayed with the 33rd. Later the 3rd Sqn was sent to the 91st Infantry Division (according to *Voigt* in September 1915; this would be a writing error meaning September 1916 as the 91st Infantry Division was only organized in September 1916) or to the 1st Landwehr Division (according to *Stein*). By 1916 all squadrons served with different divisions.

until 8 February 1917. During this period it suffered again heavy casualties. In 1916 the division contained:

33rd Infantry Division:
- 66th Infantry Brigade:
  - 98th Infantry Regiment
  - 130th Infantry Regiment
- 67th Infantry Brigade[62]:
  - 135th Infantry Regiment
  - 144th Infantry Regiment
- 33rd Field Artillery Brigade:
  - 33rd Field Artillery Regiment
  - 34th Field Artillery Regiment
- Cavalry:
  - 4/12th Jäger zu Pferd Regiment
- Engineers:
  - 1/16th Pioneer Battalion
  - ? Company, 20th Pioneer Battalion
  - 33rd Trench Mortar Company[63]
  - 33rd Divisional Bridging Train
- Signals:
  - 33rd Telephone Detachment

The division rested in the Sedan area and it returned to its old sector, at Vauquois, at the end of February 1917. On 3 May it was sent to the Champagne. It took part in the battle on that front around 20 May, suffering heavy losses. At the end of May the division returned to the Argonnes and arrived about 7 June at Vauquois. At the end of September it left the Argonnes for the Champagne. On 4 October the division arrived in the Tahure area and remained there until February 1918. By 1917 the division predominantly recruited Westphalians from the 7th Corps District. In 1917 it contained:

33rd Infantry Division:
- 66th Infantry Brigade:
  - 98th Infantry Regiment
  - 130th Infantry Regiment
  - 135th Infantry Regiment
- Cavalry:
  - 4/12th Jäger zu Pferd Regiment
- 33rd Artillery Command:
  - 283rd Field Artillery Regiment
- 132nd Pioneer Battalion:
  - 1/, 5/16th Pioneer Battalion[64]
  - 1st Reserve Company, 16th Pioneer Battalion
  - 3rd Trench Mortar Company
  - 16th Searchlight Section
- Signals:
  - 33rd Telephone Detachment
- Medical:
  - 42nd Medical Company
  - 280th Field Hospital
  - 282nd Field Hospital
  - ? Veterinary Hospital
- Transport:
  - ? Light Munitions Column

On 4 January 1918 the division went to rest and train near Sedan. On 14 March it moved to Montcornet. On 24 March the division went into the lines on the Ham-Noyon Road. By 24 March it had marched to Picardy and was heavily engaged, losing considerably on the 25th near Suzoy and Mont Renaud. It was withdrawn on 15 April and was reconstituted. It then went to the Aisne and was engaged there on 27 May. Again, on 2 June, it suffered heavy losses. From 7 June to 11 July it was rested south of Soissons. It was heavily engaged in the Second Battle of the Marne and was driven back to the Vesle. On 6 August the division moved to the Verdun area and rested until 15 August. It remained in the Verdun area until the war ended. In 1918 the division contained:

---

62 Passed on 7 October 1916 to 223rd Infantry Division and on 6 October 1918 to 44th Landwehr Division.

63 Raised in October 1915 from the 47th Heavy and one third, each, of the 118th and 137th Medium Trench Mortar Detachments and the 224th Light Trench Mortar Sections and the 33rd Light Trench Mortar Section. Distributed in autumn 1918 among the division's infantry regiments.

64 1/16th Pioneer Battalion served at the end of the war with the 333rd Pioneer Battalion.

33rd Infantry Division:
- 66th Infantry Brigade:
  - 98th Infantry Regiment
  - 130th Infantry Regiment
  - 135th Infantry Regiment
- Cavalry:
  - 4/12th Jäger zu Pferd Regiment
- 33rd Artillery Command:
  - 283rd Field Artillery Regiment
  - 76th Foot Artillery Battalion
  - 883rd Light Ammunitions Column
  - 1372nd Light Ammunitions Column
  - 1373rd Light Ammunitions Column
- 16th Pioneer Battalion:
  - 5/16th Pioneer Battalion
  - 1st Reserve Company, 16th Pioneer Battalion
  - 34th Searchlight Section
- 33rd Signal Command:
  - 33rd Telephone Detachment
- Medical:
  - 42nd Medical Company
  - 280th Field Hospital
  - 282nd Field Hospital
  - 33rd Veterinary Hospital

33rd Reserve Division: This division was originally intended as Main Reserve of the Metz Fortress and raised in the 16th Corps District. It comprised one Bavarian active infantry brigade, in peace belonging to the 2nd Bavarian Army Corps and stationed at Metz, and one Prussian reserve brigade. Still in August 1914 it advanced into France, fighting at Nomeny and marching to Verdun. It was in action on 24 and 25 August at Etain and suffered heavy losses. Through most of September the division was along the Moselle. At the end of September and in early October it returned to the Etain Region. On 8 October the 8th Bavarian Infantry Brigade attacked at Champion and Fresnes and on 9 October the 67th Reserve Regiment near Ville en Woevre. Thereafter the division took up positions near Les Eparges on the Côtes de Meuse. In November the 66th Reserve Brigade was sent to Flanders. It then returned to Lorraine on 25 November. In August 1914 the division contained:

33rd Reserve Division:
- 8th Bavarian Infantry Brigade:
  - 4th Bavarian Infantry Regiment
  - 8th Bavarian Infantry Regiment
- 66th Reserve Brigade:
  - 67th Reserve Regiment[65]
  - 130th Reserve Regiment[66]
- Attached:
  - Metz Reserve Regiment[67]
  - 2nd Reserve Machine Gun Detachment[68]
- Cavalry:
  - 2nd Reserve Hussar Regiment[69]
- Artillery[70]:
  - Ersatz Bn, 33rd Field Artillery Regiment
  - Ersatz Bn, 34th Field Artillery Regiment
  - Ersatz Bn, 69th Field Artillery Regiment
  - Ersatz Bn, 70th Field Artillery Regiment
  - 1st and 2nd Bn, 2nd Bavarian Foot Artillery Regiment
- Engineers:
  - 4/22nd Pioneer Battalion

---

65 Raised in August 1914 in the 16th Corps District in Lorraine at Metz and Saint Avold (2nd Bn). Disbanded in August 1918 and distributed among the 145th, 30th and 67th Infantry Regiments.

66 Raised in August 1914 with four battalions in the 16th Corps District in Lorraine at Metz. The 4th Bn became in September 1916 2nd Bn/273rd Reserve Regiment. Disbanded in September 1918 and distributed among the 130th, 135th and 98th Infantry Regiments.

67 This reserve regiment outside the regular wartime establishment originally consisted of four and later of two battalions. It was raised in August 1914 in the 16th Corps District surpassing the mobilization plans. It belonged at first to the Metz Fortress Main Reserve and later was handed from one formation to the next. Disbanded late in September 1914 in the Argonne Mountains and used to bring back to strength the 98th Infantry Regiment.

68 Raised still in 1914 at Metz. Originally attached to the 33rd. Reserve Division, to the Guard Cavalry Division, the 8th Army, the reinforced (King's) Own Hussar Brigade and finally to the Tongern area.

69 Raised in August 1914 by the 12th Hussar Regiment with three squadrons. Attached to the 33rd Division and the Metz Fortress Government. For 1914 the *Histories* mentions only two squadrons

The division would remain in the lines along the Meuse until the end of July 1916. From 17-31 January 1916 elements of the 66th Reserve Brigade were detached to the Bois le Pretre and suffered high losses. From February to the end of April the division fought in the battles of Combres and of Les Eparges, losing heavily. The division was reorganized in late April. In May it suffered heavy losses in the Calonne trench. On 17 July the 130th Reserve Infantry Regiment was detached to support an attack on Les Eparges. After that the division occupied a position south of Vaux les Palameix and took part in no significant actions for the rest of the year. In 1915 the division contained:

33rd Reserve Division:
- 8th Bavarian Infantry Brigade:
  - 4th Bavarian Infantry Regiment
  - 8th Bavarian Infantry Regiment
- 66th Reserve Brigade:
  - 67th Reserve Regiment
  - 130th Reserve Regiment
- Cavalry:
  - 2nd Reserve Hussar Regiment
- Artillery:
  - 33rd Reserve Field Artillery Regiment[71]
- Engineers:
  - 1st Ersatz Company, 20th Pioneer Battalion
  - 1st Landwehr Pioneer Company, 4th Corps District
  - 33rd Reserve Divisional Bridging Train
- Signals:
  - 33rd Reserve Telephone Detachment
- Attached:
  - 8th Foot Artillery Regiment (1 battery)
  - 18th Foot Artillery Regiment (elements)
  - 2/8th Reserve Foot Artillery Regiment
  - 1st & 2nd Btrys, 2nd Bn/16th Foot Artillery Regiment
  - 2nd Bn/2nd Bavarian Reserve Foot Artillery Regiment (elements)

The division left the Meuse area on 25 July 1916 and was sent to rest north of Briey. In August 1916 the 8th Bavarian Brigade and both its regiments were transferred to the newly forming 14th Bavarian Division. The division was then reorganized with the three-regiment formation. The third regiment was the 364th Infantry Regiment, which came from the 8th Ersatz Division. On 26 August the division was sent to the Verdun area, southwest of Fort Vaux. From 2 to 9 September it attacked and suffered heavy losses. From 13-21 September the division was reorganized and on 24 October it again suffered heavily by a French attack. The 2nd Bn/130th Reserve Infantry Regiment was reduced to 45 combatants. The division was withdrawn on 1 November, rested and reorganized. Its casualties were so significant that its ranks were filled with untrained men of the second Landsturm ban – that is: men over the age of 39. After these travails the division was sent to Lorraine and the Blamont sector. From January to August 1916 the division contained:

---

in this division. Half of the 2nd Sqn joined for some unspecified time the 30th Landwehr Brigade of the 5th Landwehr Division. The regiment dismounted and became the 2nd Reserve Hussar (Schützen) Regiment on 30 September 1916, the former 1/5th Reserve Uhlan Regiment joining as 4th Sqn. Disbanded on 14 May 1918.

70 The four field artillery Ersatz battalions joined in summer 1915 the new 33rd Reserve Field Artillery Regiment of three battalions.

71 Raised from the Ersatz Battalions of the 33rd, 34th, 69th and 70th Field Artillery Regiment.

33rd Reserve Division:
- 8th Bavarian Infantry Brigade[72]:
  - 4th Bavarian Infantry Regiment
  - 8th Bavarian Infantry Regiment
- 66th Reserve Brigade:
  - 67th Reserve Regiment
  - 130th Reserve Regiment
- Cavalry:
  - 2nd Reserve Hussar Regiment
  - 6/16th Uhlan Regiment[73]
- Artillery:
  - 33rd Reserve Field Artillery Regiment
- Engineers:
  - 4/16th Pioneer Battalion
  - 1st Landwehr Pioneer Company, 4th Corps District
  - 233rd Trench Mortar Company[74]
  - 33rd Reserve Divisional Bridging Train
- Signals:
  - 33rd Reserve Telephone Detachment
- Attached:
  - 4th Anti-Aircraft Detachment

The division remained in Lorraine until 10 March 1917, and after a month's rest, the division moved to the Chemin des Dames Region on 19 April. The French attack of 5 May and the subsequent counterattack inflicted heavy losses. It lost 1,000 prisoners in these and almost the entire 2nd Bn/67th Reserve Regiment was captured. In fighting on 16 May the 130th Infantry Regiment suffered heavy casualties. On 10 June the division returned to its old Blamont sector in Lorraine. Then, on 23 August it went again into action on the Verdun front. It was heavily attacked by the French on 26 August and suffered heavy losses. Leaving about the 10 September, by the beginning of October the division was in Galicia. After the Bavarians had left in 1916 the division was entirely Prussian with personnel mostly from the 7th and 8th Corps Districts. However, after the heavy losses at the Chemin des Dames young recruits of the 1918 Class were sent from West Prussia and in August and September older men of the second Landsturm ban from the 9th and 10th Corps District, who had been assembled ain the Beverloo and Warsaw depots. In 1917 the division contained:

33rd Reserve Division:
- 66th Reserve Brigade:
  - 364th Infantry Regiment[75]
  - 67th Reserve Regiment
  - 130th Reserve Regiment
- Cavalry:
  - 6/16th Uhlan Regiment[76]
- 125th Artillery Command:
  - 33rd Reserve Field Artillery Regiment
- 333rd Pioneer Battalion:
  - 1st Ersatz Company, 20th Pioneer Battalion[77]
  - 1st Landwehr Pioneer Company, 4th Corps District
  - 233rd Trench Mortar Company
  - 14th Searchlight Section
- Signals:
  - 433rd Telephone Detachment
- Medical:
  - 374th Medical Company
  - 47th Field Hospital
  - 51st Field Hospital
  - ? Veterinary Hospital
- Transport:
  - 873rd Light Munitions Column

---

72 Left on 3 August 1916 for the 14th Bavarian Infantry Division and joined on 15 September 1918 the 6th Bavarian Reserve Division.

73 At the end of 1915, according to *Stein*, the 6/16th Uhlan Regiment joined this division. At some unspecified time, though, probably in 1918, it left for the independent 31st Landwehr Brigade which occupied until February 1917 the Outer Sector IV and the Outer Sector III of the Metz Fortress.

74 Raised in October 1915 from the 56th Heavy, the 146th Medium and the 236th and 265th Light Trench Mortar Detachments. Joined in August 1918 the new Saxon 22nd Trench Mortar Battalion.

75 Combined in summer 1915 from the 32nd, 80th and 86th Brigade Ersatz Battalions of the 21st, 8th and 16th Corps Districts.

In autumn 1917 or winter 1917/1918 the division returned to France. Before the Aisne offensive in May 1918 it was back in the lines northwest of Rheims. Subsequently it marched on the Marne ending its advance near Vrigny on 31 May and was then withdrawn on 20 June. On 25 June the division went into the lines in the Mont Tetu sector in the Eastern Champagne. In the German attack of 15 July it suffered so heavily that on 20 July it was withdrawn. In mid-August the division and its infantry regiments were broken up. The 364th Infantry, the 67th Reserve and 130th Reserve Regiments were, respectively, sent to the 16th, 34th and 33rd Infantry Divisions. In January 1918 the division contained:

33rd Reserve Division:
- 66th Reserve Brigade[78]:
  - 364th Infantry Regiment
  - 67th Reserve Regiment
  - 130th Reserve Regiment
- Cavalry:
  - unknown
- 125th Artillery Command[79]:
  - 33rd Reserve Field Artillery Regiment
- 333rd Pioneer Battalion[80]:
  - 333rd Pioneer Company
  - 1st Landwehr Pioneer Company, 4th Corps District[81]
  - 14th Searchlight Section
  - 233rd Trench Mortar Company
- Signals:
  - 433rd Telephone Detachment
- Medical:
  - 273rd Medical Company
  - 51st Field Hospital
  - ? Veterinary Hospital
- Transport
  - ? Motor Transport Column
- Attached:
  - 646th Supply Depot
  - 47th Field Artillery Regiment

34th Infantry Division: The division, from the Lorraine 16th Corps District, entered France on 21 August and crossed the Meuse on 1 September at Vilosnes and Sivry. It then pushed as far as Seraucourt. After the Battle of the Marne the division withdrew and took up positions in the Argonnes. In August 1914 it contained:

34th Infantry Division:
- 68th Infantry Brigade:
  - 67th Infantry Regiment[82]
  - 145th Infantry Regiment
- 86th Infantry Brigade:
  - 30th Infantry Regiment[83]
  - 173rd Infantry Regiment
- Cavalry:
  - 1/ to 4/14th Uhlan Regiment[84]
- 34th Field Artillery Brigade:
  - 69th Field Artillery Regiment
  - 70th Field Artillery Regiment
- Support Troops:
  - 2/, 3/16th Pioneer Battalion
  - 34th Divisional Bridging Train
  - 2nd Medical Company

---

76 The *Histories* attributes the 4/19th Hussar Regiment to this division during 1916. According to *Stein*, this served with the 19th Ersatz Division and later with the 219th Infantry Division.

77 This company, according to *Stein*, served with the 333rd Pioneer Battalion towards the end of the war. Also, for some time, the Saxon 333rd Pioneer (Miner) Company would be assigned for some unspecified time to this division before joining the 1st Army.

78 Disbanded on 1 September 1918.

79 Left for the 3rd Infantry Division on 12 October 1918.

80 Left for the Gallwitz Army Group.

81 Left for the independent 61st Landwehr Brigade.

82 A regiment seconded out of area from the 4th Corps District.

83 A regiment seconded out of area from the 8th Corps District.

84 The *Histories* attributes in 1914 the 16th Uhlans. However its two half regiments were then serving with the 13th and 14th Infantry Divisions. According to *Stein*, instead the 14th Uhlan

The division served in the Argonne, without interruption, from September 1914 to 15 August 1916. In January and July 1915 it participated in two offensives and suffered heavy losses. By 18 January the 30th Infantry Regiment had lost 56 officers and 2,723 men. It received heavy replacements after these actions. Its losses after that, until 15 August 1916, were minimal. In 1915 the division contained:

34th Infantry Division:
- 68th Infantry Brigade:
  - 67th Infantry Regiment
  - 145th Infantry Regiment
- 86th Infantry Brigade:
  - 30th Infantry Regiment
  - 173rd Infantry Regiment
- Cavalry:
  - 2/, 3/12th Jäger zu Pferd Regiment[85]
- 34th Field Artillery Brigade:
  - 69th Field Artillery Regiment
  - 70th Field Artillery Regiment
- Engineers:
  - ? Company, 16th Pioneer Battalion
  - 34th Divisional Bridging Train
- Signals:
  - 34th Telephone Detachment

On 15 August 1916 the division was sent to Verdun. In late August and early September the division fought at Chiaumont. On 20 September and again on 24 October it suffered heavy losses when the French recaptured Fort Douaumont and the Thiaumont Works. On 29 October the division was relieved and sent to a quiet sector in the Vosges around Ssenonces. It received considerable reinforcements at this time and the 173rd Infantry Regiment was transferred to the new 223rd Division. In 1916 the division contained:

34th Infantry Division:
- 68th Infantry Brigade:
  - 67th Infantry Regiment
  - 145th Infantry Regiment
- 86th Infantry Brigade[86]:
  - 30th Infantry Regiment
  - 173rd Infantry Regiment
- 34th Field Artillery Brigade:
  - 69th Field Artillery Regiment
  - 70th Field Artillery Regiment
- Cavalry:
  - 5/12th Jäger zu Pferd[87]
- Engineers:
  - 2/16th Pioneer Battalion
  - 34th Trench Mortar Company[88]
  - 34th Divisional Bridging Train
- Signals:
  - 34th Telephone Detachment
- Attached:
  - Labor Battalion, 34th Division

Regiment was present. It was mobilized in August 1914 leaving the 5th Sqn at home at Hanover, 10th Corps District, as Ersatz Sqn. It apparently left in early 1915 for the 5th Reserve Corps and on 2 August 1916 was broken up. According to *Voigt* the staff, 1st and 4th Sqns joined the 199th Infantry Division. The 4th left and after a spell (according to *Stein*) with the 105th in late 1916 joined the 220th Infantry Division on 16 January 1917. The 2nd Sqn, after a spell (according to *Stein*) with the 195th, joined the 197th Infantry Division. Finally the 3rd Sqn was attached at first to the 197th and subsequently to the 48th Landwehr Division, apparently after the 197th had been disbanded.

85 According to *Stein*, those two squadrons left the 33rd Infantry Division and were at first attached to the 34th and subsequently to the 121st Infantry Divisions. As the latter was raised already in March 1915 their stay with the 34th was a short one. Later the 5/12th Jäger zu Pferd Regiment also joined from the 33rd Infantry Division replacing the other two squadrons.

86 Disbanded on 3 November 1916.

87 Joined still in 1915.

88 Raised in October 1915 from the 8th Heavy, the 138th Medium and the 205th Light Trench Mortar Detachments. Distributed in autumn 1918 among the division's infantry regiments.

In February 1917 the division was sent to the Argonnes and in mid-March to a position north of Rheims. The 67th and 30th Infantry Regiments went into action near Brimont, in the Champagne, while the 145th Infantry Regiment was sent to Cornillet to reinforce that sector in mid-April. About 25 April the division was concentrated near Brimont and in mid-June it suffered heavy losses. The 15th Infantry lost about one third of its effectives. On 21 July the division was sent to Flanders, arriving on 7 August. On 12 August it went into action and suffered heavily. On 24 August it was sent to the Woevre Region, southeast of Thiaucourt and remained there until 31 October. Subsequently it was sent to Cambrai and participated in the German counterattack on 30 November. After a short rest, during early December, the division went to the Laon Region. It remained on the Ailette until 7 January 1918. Because of the recruiting difficulties in Lorraine, the division consisted in 1917 predominantly of personnel from the 7th and 8th Corps Districts. In September 1917 large numbers of men over the age of 25 and Landsturm personnel over 30 were taken in who had been drafted from divisions on the Eastern Front and retrained for some time in the interior of Germany. In 1917 the division contained:

34th Infantry Division:
- 68th Infantry Brigade:
  - 30th Infantry Regiment
  - 67th Infantry Regiment
  - 145th Infantry Regiment
- Cavalry:
  - 5/12th Jäger zu Pferd
- 34th Artillery Command:
  - 70th Field Artillery Regiment
- 132nd Pioneer Battalion:
  - 2/, 3/16th Pioneer Battalion
  - 34th Trench Mortar Company
  - 259th Searchlight Section
  - 34th Divisional Bridging Train
- Signals:
  - 34th Telephone Detachment
- Medical:
  - Medical Company
  - 281st Field Hospital
  - 282nd Field Hospital
  - 283rd Field Hospital
  - ? Veterinary Hospital
- Transport:
  - ? Light Munitions Column

After resting near Laon, the division went into the lines near Courtecon. It was withdrawn in late February and rested near Liesse until 15 March. At that point the division went to Picardy and took part in the German offensive. On 2 April it was withdrawn, after having lost 50% of its effectives. It received 1,000 replacements shortly afterwards. The division remained in Picardy until early June, when it went to the Aisne Region. It attacked on 12 June, suffering 30% of its strength as casualties and was withdrawn on 6 July. The division went back to the lines from 19 to 22 July and then was withdrawn, having lost heavily yet again. About this time it received drafts from the dissolved 67th Reserve Regiment of the disbanded 33rd Reserve Division. After a short period in the Ailette Region, on 28 September the division was near St. Quentin. It was forced out of St. Quentin by the French on 30 September. It was then sent to the Guise Region and went into the lines almost immediately upon arriving. The division was savaged, losing more than 700 prisoners. It was sent to the Vervins Region to rest. On 1 November it returned to the lines near Puisieux and remained there until the armistice. In 1918 the division contained:

34th Infantry Division:
- 68th Infantry Brigade:
  - 30th Infantry Regiment
  - 67th Infantry Regiment
  - 145th Infantry Regiment
- Cavalry:
  - 5/12th Jäger zu Pferd

34th Artillery Command:
- (?) Field Artillery Regiment
- 116th Foot Artillery Battalion
- 1174th Light Ammunitions Column
- 1191st Light Ammunitions Column
- 1192nd Light Ammunitions Column

132nd Pioneer Battalion:
- 2/, 3/16th Pioneer Battalion
- 34th Trench Mortar Company
- 193rd Searchlight Section

34th Signal Command:
- 34th Telephone Detachment

Medical:
- 41st Medical Company
- 281st Field Hospital
- 283rd Field Hospital
- 34th Veterinary Hospital

Transport:
- 567th Motor Transport Column

Attached (Elements attached 6 June 1918):
- 44th Machine Gun Sharpshooter Battalion
- 249th Reconnaissance Flight
- 119th Artillery Observation Section
- 83rd Balloon Squadron
- 534th Carrier Pigeon Loft

35th Infantry Division: When the war broke out the 35th Division of the 17th Corps District went to the Eastern Front. It fought at Gumbinnen and Tannenberg and participated in two offensives on Warsaw. It then participated in the offensives on Bzura and Narew, where it remained until 1 August 1915. In August 1914 the division contained:

35th Infantry Division:
- 70th Infantry Brigade:
  - 21st Infantry Regiment
  - 61st Infantry Regiment
- 87th Infantry Brigade:
  - 141st Infantry Regiment
  - 176th Infantry Regiment
  - 2nd Jäger Battalion

Cavalry:
- 1/, 2/, 4/, 5/4th Jäger zu Pferd Regiment[89]

35th Field Artillery Brigade:
- 71st Field Artillery Regiment
- 81st Field Artillery Regiment

Support Troops:
- 1/17th Pioneer Battalion
- 35th Divisional Bridging Train
- 1st & 3rd Medical Companies

In August 1915 the division was sent to rest at Bialistok. On 6 October 1915 the division was sent to the Western Front. In mid-October it arrived in the Roye sector and remained there through the winter of 1915/1916. In January 1915 the division contained:

---

[89] Mobilized in August 1914 leaving the 3rd Sqn as its Ersatz Sqn at Graudenz, 17th Corps District (mobilized itself as part of the Graudenz Fortress Garrison on 14 December 1914; joined the Breugel Division, became the 2nd Mobile Ersatz Sqn, 17th Corps District, and finally the 3/85th Cavalry Regiment; a new Ersatz Sqn was raised on 14 December 1914 by the 17th Corps District). Already in February 1915 the 2nd Sqn together with the 4/5th Hussar Regiment (from the 36th Infantry Division) and the 1/ and 2/2nd Dragoon Regiment (from the 2nd Cavalry Division; however, it could be that there is a printing error and *Stein* actually means the 1/ and 2/2nd Reserve Dragoon Regiment from the 5th Reserve Division, which formed a Tschammer Regiment raised in early February 1915), joined the improvised Borcke Cavalry Regiment, which was formed by part of the previous Tschammer Regiment. *Stein* does not tell how long that seconding lasted. The 2nd subsequently joined the 35th Reserve Division.

So, apparently, the 35th Infantry Division just retained 1st, 4th and 5th Sqns of the regiment during 1915 and early 1916. From summer 1916 onwards the divisional cavalry was to be reduced to one squadron. The 1st Sqn joined the 226th Infantry Division raised in November 1916 and when that was disbanded in May 1918 the Guard Cavalry Schützen Division. The 4th left in autumn 1916 for the 47th Reserve Division. The 5th Sqn joined the 105th Infantry Division.

35th Infantry Division:
70th Infantry Brigade[90]:
21st Infantry Regiment
61st Infantry Regiment
87th Infantry Brigade:
141st Infantry Regiment
176th Infantry Regiment
Cavalry:
4th Jäger zu Pferd Regiment
35th Field Artillery Brigade:
71st Field Artillery Regiment
81st Field Artillery Regiment
Engineers:
1/17th Pioneer Battalion
35th Divisional Bridging Train
Signals:
35th Telephone Detachment

May 1916 found the division on the Somme. In July, when the Somme battle began the division was west of Vermandovillers and south of Chilly. The 176th Infantry Regiment was in the Herbécourt-Estrées sector on 2 July. It again suffered heavy casualties in a French attack on 20 July. On 4 September the division was in the process of moving into the lines when it was struck by a French offensive and lost 2,000 prisoners and more casualties. The division was withdrawn from the front on 8 September and rested at Ham. It was back in the lines from 15 to 20 October southwest of Chaulnes. Official records show it had lost 6,102 men casualties, 68% of its effectives, during the course of the Battle of the Somme. In 1916 the division contained:

35th Infantry Division:
87th Infantry Brigade:
61st Infantry Regiment
141st Infantry Regiment
176th Infantry Regiment
Cavalry:
1/, 4, 5/4th Jäger zu Pferd Regiment
2/5th Hussar Regiment[91]
35th Field Artillery Brigade:
71st Field Artillery Regiment
81st Field Artillery Regiment
Engineers:
1/17th Pioneer Battalion
4/36th Pioneer Battalion
35th Trench Mortar Company[92]
35th Divisional Bridging Train
Signals:
35th Telephone Detachment
Attached:
2nd Assault Detachment

In March 1917, when the Germans retired to the Hindenburg Line, the division retreated into a position south of St. Quentin. After resting for a few days at the beginning of April, the division participated in the Battle of Arras in late April. By beginning of May it had lost about 50% of its effectives. The 141st Infantry Regiment was rebuilt using men of the Class of 1918 and men from the disbanded 618th Infantry Regiment. The division rested all of May and was rebuilt with 3,000 replacements from the Warsaw recruit depot. It was then sent to Flanders and on 7 June it lost heavily on the banks of the Ypres-Comines Canal. The division's losses are estimated at from 5,000 to 6,000 men, of whom 1,272 were prisoners. It then was reorganized around 11 June near Cambrai with returned convalescents and wounded. It was subsequently sent to a calm sector north of St. Quentin remaining there from 21 June to 20 October. On 22 October it arrived in Flanders. It suffered serious losses from 22 to 25 October, but was relieved only on 22 January 1918, when it was sent to rest at Bruges. In 1917 the division still

90 Disbanded on 15 May 1915.

91 For 1916 and 1917 the *Histories* attributes two squadrons of this hussar regiment, which according to *Stein* is not correct. According to *Voigt*, the squadron joined on 18 November 1915 for the remainder of the war.

92 Raised in November 1915 from the 163rd Medium Trench Mortar Detachment and the 7th Heavy and one quarter of the 113th Light Trench Mortar Section.

consisted of personnel from its home West Prussian 17th Corps District. However, there were also drafts from the Silesian 5th and 6th Corps Districts, especially miners drafted in June 1916. Quite a number of the men, about 50% of the 141st Infantry Regiment, were Polish speaking. In 1917 the division contained:

35th Infantry Division:
- 87th Infantry Brigade:
  - 61st Infantry Regiment
  - 141st Infantry Regiment
  - 176th Infantry Regiment
- 133rd Pioneer Battalion:
  - 1/, 2/17th Pioneer Battalion
  - 35th Trench Mortar Company
  - 35th Divisional Bridging Train
- Signals:
  - 35th Telephone Detachment
- Cavalry:
  - 2/5th Hussar Regiment
  - 4/14th Hussar Regiment[93]
- 35th Artillery Command:
  - 71st Field Artillery Regiment
- Medical:
  - 44th Medical Company
  - 289th Field Hospital
  - 293rd Field Hospital
  - 35th Veterinary Hospital
- Transport:
  - 568th Motor Transport Column

From 17 February to 20 March 1918 the division was engaged near Merckem. It was withdrawn then and reentered the lines near Lille where it was engaged on 8/9 April in the Battle of the Lys. In further fighting on 12 April it suffered heavily. On 5 May the division went to the sector north of Bailleul and remained there from 8 May to 3 July. During this period the effective strength of its infantry companies was about 50 men. On 17 May the division went back into the lines in the Merckem sector and remained there until 18 August. During this period it received substantial reinforcements.

After resting near Lille on 25 August the division went to Douai. It went into the Drocourt-Queant line and fought there until about 30 September. It was relieved after suffering heavy casualties and about 800 prisoners. On 1 October the division was engaged in the Battle of Cambrai. It began withdrawing as the war wound down and on 9 November was near Harchies. In 1918 the division contained:

35th Infantry Division:
- 87th Infantry Brigade:
  - 61st Infantry Regiment
  - 141st Infantry Regiment
  - 176th Infantry Regiment
- Cavalry:
  - 2/5th Hussar Regiment
- 35th Artillery Command:
  - 71st Field Artillery Regiment
  - 1st Bn/18th Reserve Foot Artillery Regiment
  - 715th Light Ammunitions Column
  - 716th Light Ammunitions Column
  - 1305th Light Ammunitions Column
- 133rd Pioneer Battalion:
  - 1/, 2/17th 17th Pioneer Battalion
  - 35th Trench Mortar Company
  - 223rd Searchlight Section
- 35th Signal Command:
  - 35th Telephone Detachment
  - 121st Wireless Detachment
- Medical:
  - 44th Medical Company
  - 289th Field Hospital
  - 293rd Field Hospital
  - 35th Veterinary Hospital
- Transport:
  - 568th Motor Transport Column

35th Reserve Division: The 35th Division started the war as the Main Reserve of the Thorn Fortress and remained in the east throughout the war. Between 25 and 28 August 1914 it fought on the southern frontier of East Prussia. Subsequently, it took part in the

---

93 Joined from 7 March to 25 September 1917 according to *Voigt* coming from the 2nd Cavalry Division and returning to it again.

Polish campaign. In October, it was southwest of Warsaw and in December near Czenstochow. In August 1914 the division contained:

35th Reserve Division:
- 5th Landwehr Infantry Brigade:
  - 2nd Landwehr Regiment[94]
  - 9th Landwehr Regiment[95]
- 20th Landwehr Brigade:
  - 19th Landwehr Regiment[96]
  - 107th Saxon Landwehr Regiment[97]
- Attached:
  - 4th Reserve Machine Gun Detachment[98]
  - 1st, 2nd, 3rd & 4th Fortress Machine Gun Detachments

Cavalry:
- 3rd Heavy Reserve Cavalry Regiment[99]

Artillery[100]:
- Ersatz Battalion, 35th Field Artillery Regt
- Ersatz Battalion, 81st Field Artillery Regt
- 1st Bn/11th Reserve Foot Artillery Regiment (3/4)
- 4/15th Reserve Foot Artillery Regiment (100mm)

Engineers:
- 1st Reserve Company, 17th Pioneer Battalion

From January to March 1915, the division was in line south of Pilica. From April to June, it fought in the Carpathians, and later near Lemberg. In July it was relieved in Galicia by the 39th Austrian Division and, until October 1915, it pursued Russian forces from Cholm via Brest-Litovsk and Kobrin to the Oginski Canal. There it took up positions near Logishin and remained there for the next two years until February 1918. In 1915 the division consisted of:

---

94 Raised in August 1914 in the 2nd Corps District by the 42nd Infantry (at Stralsund) and the 2nd Grenadier Regiments (3rd Bn; at Stettin). The active 3rd Fortress Machine Gun Detachment was attached and became the regimental machine gun company. Joined the 226th Infantry Division, in November 1916. The regiment was especially honored in July 1917 for the defensive battle at Smorgon-Krevo to be called Landwehr Infantry Regiment "König Wilhelm II" No. 2 – unique for a Landwehr Regiment.

95 Raised in August 1914 in the 2nd Corps District by the 9th Grenadier (at Stargard) and the 54th Infantry Regiments (3rd Bn; at Colberg). Disbanded in May 1918 and distributed among the 4th Infantry Division.

96 Raised in August 1914 in the 5th Corps District in Silesia by the 58th Infantry (at Glogau) and the 7th Grenadier Regiments (3rd Bn; at Liegnitz). The 1st Fortress Machine Gun Detachment (raised from the active 5th Fortress Machine Gun Company) was attached and became the regimental machine gun company. A 4th Bn was raised later, sent to East Prussia in spring 1915 and became 3rd Bn/381st Infantry Regiment in summer 1915. Regiment joined on 28 March 1915 the 4th Landwehr Division, later the 18th Landwehr Division and in October 1916 the 224th Infantry Division.

97 Raised in August 1914 in the 19th Corps District by the 133rd Infantry Regiment at Zwickau. Joined 45th Landwehr Division in February 1917.

98 Raised still in 1914 and at first attached to the 35th Reserve Division. Joined subsequently the Alpine Corps, the 200th Infantry Division, the "A" Army Detachment and the 301st Infantry Division.

99 Raised in August 1914 by the 3rd Cuirassier Regiment with three squadrons. Dismounted on 9 September 1916 and converted into the 3rd Heavy Reserve Cavalry (Schützen) Regiment including the former 1st Landwehr Sqn, 8th Corps District, and the Machine Gun Sqn/13th Uhlan Regiment. Disbanded on 12 April 1918 and mostly versed into the Field Recruit Depot of the 41st Infantry Division.

100 The two field artillery Ersatz battalions already at Tannenberg in August 1914 formed the improvised Wittich Ersatz Field Artillery Regiment, which in summer 1915 became the 35th Reserve Field Artillery Regiment.

35th Infantry Reserve Division:
- 5th Landwehr Infantry Brigade:
  - 2nd Landwehr Regiment
  - 9th Landwehr Regiment
- 20th Landwehr Brigade[101]:
  - 19th Landwehr Regiment
  - 107th Landwehr Regiment
- Cavalry:
  - 3rd Heavy Reserve Cavalry Regiment
- Artillery:
  - 35th Reserve Field Artillery Regiment[102]
- Engineers[103]:
  - 2/17th Pioneer Battalion

In October 1916, the 5th Landwehr Infantry Brigade and its two regiments left for the new 226th Division forming in the Smorgoni sector. They were replaced by the 167th Brigade[104] with the new 420th and 421st Infantry Regiments. Before that, in 1916, the division comprised:

35th Reserve Division:
- 5th Landwehr Infantry Brigade[105]:
  - 2nd Landwehr Regiment
  - 9th Landwehr Regiment
  - 107th Landwehr Regiment
- Cavalry:
  - 3rd Heavy Reserve Cavalry Regiment
- Artillery:
  - 35th Reserve Field Artillery Regiment
  - 833rd Battery
- Engineers:
  - 235th Trench Mortar Company[106]
- Attached:
  - 5th Landsturm Pioneer Company, 9th Corps District[107]

In an effort to sort out Prussian and Saxon troops, in July 1917, the Baden 438th Infantry Regiment joined, replacing the Saxon 107th Landwehr Regiment. This regiment was transferred to the Saxon 45th Landwehr Division. In 1917 the division contained:

35th Reserve Division:
- 167th Infantry Brigade:
  - 420th Infantry Regiment[108]
  - 421st Infantry Regiment[109]
  - 438th Infantry Regiment[110]
- Cavalry:
  - 2/4th Jäger zu Pferd Regiment[111]
- Artillery:
  - 35th Reserve Field Artillery Regiment
- 335th Pioneer Battalion:
  - 1st Reserve Company, 17th Pioneer Battalion
  - 235th Trench Mortar Company
  - 35th Reserve Searchlight Section
- Signals:
  - ? Telephone Detachment
- Medical:
  - 238th Medical Company
  - ? Field Hospital
  - ? Veterinary Hospital
- Transport:
  - ? Light Munitions Column
- Attached:
  - 93rd Reserve Brigade, reinforced by 20th & 24th Landsturm Battalions

---

101 Disbanded on 16 July 1915.

102 Raised in summer 1915 including the 35th and 81st Field Artillery Ersatz Battalions.

103 According to *Stein*, also the 1st Landsturm Pioneer Company, 5th Corps District served with this division before joining the 10th Army.

104 Originally, the 12th Ersatz Brigade with the 84th Infantry Division. On 12 June 1915 renamed as 167th Infantry Brigade. Joined the 35th Reserve Division on 24 November 1916.

105 Joined, on 2 December 1916, the 226th Infantry Division and, on 25 April 1918, the Guard Cavalry Schützen Division (the dismounted Guard Cavalry Division).

106 Raised in October 1915 from the 176rd Heavy and the 248th Medium Trench Mortar Detachments and the 70th Trench Mortar Reinforcement. Joined in August 1918 the new 17th Trench Mortar Battalion.

107 Subsequently joined the 38th Landwehr Division.

108 Raised on 1 October 1916 by the Supreme Commander East from the Ersatz Battalion, 47th Reserve Division, the Novaya Mysh Ersatz Battalion, 3rd Landwehr Division, and the 8th Recruit

In March 1918 the division advanced into the Ukraine after sending drafts to the 10th Landwehr Division leaving for the Western Front. In April 1918, it was in the Gomel area. Apparently, the division remained there until the end of the war. Reports of sighting the division in Flanders in September would be faulty. In 1918, the division contained:

35th Reserve Division:
  167th Infantry Brigade:
    420th Infantry Regiment
    438th Infantry Regiment
  Cavalry:
    2/4th Jäger zu Pferd Regiment
  Artillery:
    unknown
335th Pioneer Battalion:
  1st Reserve Company, 17th Pioneer Battalion
  149th Searchlight Section
435th Signal Command:
  435th Telephone Detachment
Medical:
  238th Medical Company
  291st Field Hospital
  435th Veterinary Hospital

36th Infantry Division: In August 1914 the 36th Division from the West Prussian 17th Corps District went to East Prussia and came under Hindenburg's 8th Army. It fought at Tannenberg on 30 August, Lötzen on 9 September and with Mackensen's 9th Army at Radom on 6 October. Subsequently it advanced on Warsaw. In the subsequent retreat its regiments, especially the 5th Grenadiers, suffered considerable losses, principally at Lodz between 23 November and 6 December. In August 1914 the division contained:

36th Infantry Division:
  69th Infantry Brigade:
    129th Infantry Regiment
    175th Infantry Regiment
  71st Infantry Brigade:
    5th Grenadier Regiment
    128th Infantry Regiment
Cavalry:
  1/ to 4/5th Hussar Regiment[112]
36th Field Artillery Brigade:
  36th Field Artillery Regiment
  72nd Field Artillery Regiment
Support Troops:
  2/, 3/17th Pioneer Battalion
  36th Divisional Bridging Train
  2nd Medical Company

Battalion, 12th Army. Replacements were sent from the 2nd Ersatz Bn/43rd Infantry Regiment (from the 1st Corps District).

109 Raised on 1 October 1916 by the Supreme Commander East, originally as Haliersch Infantry Regiment. Replacements were sent from the 1st Ersatz Bn/140th Infantry Regiment (from the 2nd Corps District).

110 Raised on 1 October 1916 by the Supreme Commander East from drafts on Baden regiments. Replacements were sent from the (2nd) Ersatz Bn/112th Infantry Regiment (from the 14th Corps District).

111 Joined in autumn 1916.

112 Mobilized in August 1914 leaving the 5th Sqn at home at Stolp, Pomerania, 2nd Corps District, as Ersatz Sqn. Already in December 1914 the Staff and 1st Sqn together with the 1/ and 2/6th Reserve Dragoon Regiment (from the 19th Reserve Division) formed the improvised Keyser Cavalry Regiment of the 7th Army while in February 1915 the 4th Sqn together with the 2/4th Jäger zu Pferd Regiment (from the 35th Infantry Division) and the 1/ and 2/2nd Dragoons (or 2nd Reserve Dragoons ?) joined the improvised Borcke Cavalry Regiment. It is not known when exactly those squadrons returned to the division after the improvised regiments were dissolved again at some time not mentioned by *Stein*.

According to *Voigt* the 1st Sqn served with the 36th Infantry Division in 1915/1916 and joined the 15th Landwehr Division from 28 June 1916 to 26 March 1917. Until 1919 it stayed in the Ukraine. The 2nd Sqn left the division in 1915 for a spell with the Bacmeister Landsturm

During the winter of 1914/15 the division fought along the Bzura and continued there until June. In July it was on the Narew, later on the Bug and in early September on the Chtchara River. At the end of September 1915, however, it was shipped to the Western Front. The division went into the lines on 16 October in the Roye sector, but it was not seriously engaged until the Battle of the Somme. In 1915 it contained:

36th Infantry Division:
- 69th Infantry Brigade:
  - 129th Infantry Regiment
  - 175th Infantry Regiment
- 71st Infantry Brigade[113]:
  - 5th Grenadier Regiment
  - 128th Infantry Regiment
- Cavalry:
  - 5th Hussar Regiment (parts)
- 36th Field Artillery Brigade:
  - 36th Field Artillery Regiment
  - 72nd Field Artillery Regiment
- Engineers:
  - 3/17 Pioneer Battalion
  - 36th Divisional Bridging Train
- Signals:
  - 36th Telephone Detachment

In May 1916 the 129th Infantry Regiment was detached and sent to the newly forming 105th Division. From June onwards only the 4/5th Hussar Regiment was left with the 36th Division in 1916. The division remained on the Somme and on the Allied July offensive, it was south of Chilly. It was not engaged, as a whole, until October. Prior to that point it had sent elements to support other sectors that were under heavy attack. The division moved about the Somme until 24 to 25 October when it went into the lines south of the Ablaincourt-Chaulnes Wood and came under heavy attack. It suffered considerably, especially the 128th Infantry Regiment. In 1916 the division contained:

36th Infantry Division:
- 71st Infantry Brigade:
  - 5th Grenadier Regiment
  - 128th Infantry Regiment
  - 175th Infantry Regiment
- Cavalry:
  - 4/5th Hussar Regiment
- 36th Field Artillery Brigade:
  - 36th Field Artillery Regiment
  - 72nd Field Artillery Regiment
- Engineers:
  - 3/17th Pioneer Battalion
  - 36th Trench Mortar Company[114]
  - 36th Divisional Bridging Train
- Signals:
  - 36th Telephone Detachment
- Attached Gas Troops:
  - 3/, 4/35th Pioneer Regiment
  - 3/36th Pioneer Regiment

In March 1917 the division retired to the Hindenburg line in St. Quentin area. It remained there from 23 March until 9 April, when it rested. On 9 May it moved into the Artois Region and did not participate in any serious engagements. In June it rested and on 4 July was in the Oppy-Gavrelle sector. It was relieved in August and sent to the Ypres Region, where, near Poelcapelle it suffered a heavy British attack. Subsequently the division was withdrawn from the lines and sent to a calm sector near St. Quentin where it remained until February 1918. In 1917 the division contained:

Detachment and joined the 35th Infantry Division on 18 November 1915 and stayed there until the end of the war. The 4th Sqn served with the 36th Infantry Division from sometime in 1915 after the 2nd had left until the war's end. The 3rd served from 12 May to 22 November 1915 with the 107th Infantry Division. At a date not mentioned by *Stein*, it joined the 11th Army, the 302nd Infantry Division, the Quiring Detachment, the Alpine Corps and finally the 6th Reserve Division.

113 On 15 May 1915 renamed as 209th Infantry Brigade of the 105th Infantry Division.

114 Raised in April 1916 from the 45th Heavy, the Medium Guard and the 267th Light Trench Mortar Sections.

36th Infantry Division:
- 71st Infantry Brigade:
  - 5th Grenadier Regiment
  - 128th Infantry Regiment
  - 175th Infantry Regiment
- Cavalry:
  - 4/5th Hussar Regiment
- 36th Artillery Command:
  - 36th Field Artillery Regiment
- Engineers:
  - 3/, 5/17th Pioneer Battalion
  - 36th Trench Mortar Company
- Signals:
  - 36th Telephone Detachment
- Medical:
  - 43rd Medical Company
  - 288th Field Hospital
  - 290th Field Hospital
  - 36th Veterinary Hospital

About 1 February the division was north of St. Quentin and moved into the lines south of there a few days later. It was in the lines during the Somme Offensive (Kaiserschlacht), advancing as far as Candor in Picardy. The division rested from 8 to 20 April. From 20 to 28 April it was engaged near Montdidier. It received 1,000 replacements on 29 April. The division rested from 9 to 23 May and was engaged in the Battle of the Aisne from 27 May until mid-June. On 30 June it withdrew from the sector east of Chateau Thierry. It rested until 14 July and was rebuilt during that period. On 15 July it was engaged, crossing the Marne and penetrating as far as Chartevos. It then fought at Rocourt and Villeneuve sur Fore until 27 July. It rested near Laon in early August and moved forward to resist the British attack north of Bapaume on 24 August. Under-strength, it could not resist and was driven back to Inchy. It lost heavily, including 800 prisoners. It continued in the lines, retreating to Provin on 16 September and it was last identified near Bany on 10 November. In September 1918 the division received the men from the 377th Infantry Regiment when the 10th Landwehr Division was dissolved. In 1918 the division contained:

36th Infantry Division:
- 71st Infantry Brigade:
  - 5th Grenadier Regiment
  - 128th Infantry Regiment
  - 175th Infantry Regiment
- Cavalry:
  - 4/5th Hussar Regiment
- 36th Artillery Command:
  - 36th Field Artillery Regiment
  - 1st Bn/4th Reserve Foot Artillery Regiment
  - 824th Light Ammunition Column
  - 1209th Light Ammunition Column
  - 1229th Light Ammunition Column
- 17th Pioneer Battalion:
  - 3/, 5/17th Pioneer Battalion
  - 36th Trench Mortar Company
  - 206th Searchlight Section
- 36th Signal Command:
  - 36th Telephone Detachment
  - 62nd Wireless Detachment
- Medical & Transport:
  - 43rd Medical Company
  - 288th Field Hospital
  - 290th Field Hospital
  - 36th Veterinary Hospital
  - 569th Motor Transport Column

36th Reserve Division: At the beginning of the war the 36th Reserve Division, raised in the 17th Corps District, was in East Prussia and it was quickly engaged near Gumbinnen and Angerburg. The division had one active infantry regiment from the 2nd Army Corps. It served under Mackensen and on 6 November was fighting on the left bank of the Vistula and in early December on the Bzura. In August 1914 the division contained:

36th Reserve Division:
69th Reserve Brigade:
21st Reserve Regiment[115]
61st Reserve Regiment[116]
2nd Reserve Jäger Battalion[117]
70th Reserve Brigade:
54th Infantry Regiment
5th Reserve Regiment[118]

Cavalry:
1st Reserve Hussar Regiment[119]
Artillery:
36th Reserve Field Artillery Regiment (6 btrys)[120]
Engineers:
1st Reserve Company, 2nd Pioneer Battalion
36th Reserve Divisional Bridging Train
15th Reserve Medical Company

In early 1915 the division fought along the Bzura-Rawka-Bolimov line. On 13 February it moved to the right wing of the German Army, which was driving the Russians out of East Prussia. In April it attacked near Prasnysz, but suffered heavy losses to Russian counterattacks. In May it fought in the Hindenburg Offensive in Courland and by September was near Dvinsk. It suffered heavy losses during this advance. The 61st Reserve Regiment had an average of 80 men per company at this time. In December the division was in the lines southeast of Riga. In 1915 the division contained:

36th Reserve Division:
69th Reserve Brigade:
21st Reserve Regiment
61st Reserve Regiment
70th Reserve Brigade:
54th Infantry Regiment
5th Reserve Regiment
Attached:
2nd Reserve Jäger Battalion

Cavalry:
1st Reserve Hussar Regiment
Artillery:
36th Reserve Field Artillery Regiment (7 btrys)
Engineers:
1st Reserve Company, 2nd Pioneer Battalion
2/32nd Reserve Pioneer Battalion
36th Reserve Divisional Bridging Train
Signals:
36th Reserve Telephone Detachment

The division remained in Courland, near Riga until 24 September 1916. In late September or early October it was transferred to Galicia. The 54th Infantry Regiment was engaged on 3 October against advancing Russians. This regiment was rejoined by the rest of the division on 19 October and remained near Brzezany. In 1916 the division contained:

36th Reserve Division:
69th Reserve Brigade:
21st Reserve Regiment
61st Reserve Regiment

70th Reserve Brigade[121]:
54th Infantry Regiment
5th Reserve Regiment

---

115 Raised in August 1914 in the 17th Corps District at Thorn (1st Bn), Graudenz (Staff, 2nd Bn) and Konitz (3rd Bn). Joined the 216th Infantry Division in September 1917 and later the 217th.

116 Raised in August 1914 in the 17th Corps District at Danzig and Neustadt in West Prussia (2nd Bn).

117 Raised in August 1914 in the 17th Corps District by the 2nd Jäger Battalion at Kulm. No machine gun company. Joined the 12th Jäger Regiment in October 1917.

118 Raised in August 1914 in the 17th Corps District at Schlawe (1st Bn), Stolp (Staff, 2nd Bn) and Preussisch Stargard (3rd Bn).

119 Raised in August 1914 with three squadrons by the 5th Hussar Regiment. At some date, not provided by *Stein*, it as seconded to the improvised Berner Cavalry Brigade. On 16 August 1916 the regiment dismounted and became the 1st Reserve Hussar (Schützen) Regiment incorporating as 4th Sqn the former 76th Reserve Cavalry Detachment and a machine gun squadron. It was disbanded on 12 April 1918.

120 Raised in August 1914 at Allenstein by the 73rd Field Artillery Regiment.

Cavalry:
1/1st Reserve Uhlan Regiment[122]
Artillery:
36th Reserve Field Artillery Regiment (7 btrys)
Engineers:
1st Reserve Company, 2nd Pioneer Battalion
236th Trench Mortar Company[123]
36th Reserve Divisional Bridging Train
Signals:
36th Reserve Telephone Detachment
Attached:
80th Artillery Survey Section
24th Sound Ranging Section
217th Anti-Aircraft Detachment

At the end of May 1917 the division was sent to the Western Front, arriving in Lorraine on 1 June. It was trained until 24 June and then went into the Haye sector for two weeks. In July it was sent to a position near Lens, in the Artois, occupying the Mericourt sector until early October. On 20 October it was sent to Flanders and went into the lines near Ypres. Though recruited in West Prussia and Eastern Pomerania, the division received a large number of drafts from Alsace-Lorraine after it had moved to the Western Front. In 1917 it contained:

36th Reserve Division:
69th Reserve Brigade:
54th Infantry Regiment
5th Reserve Regiment
61st Reserve Regiment
Cavalry:
5/2nd Guard Dragoon Regiment[124]
72nd Artillery Command:
36th Reserve Field Artillery Regiment (9 btrys)
336th Pioneer Battalion[125]:
1st Reserve Company, 2nd Pioneer Battalion
236th Trench Mortar Company
Signals:
436th Telephone Detachment
Medical:
515th Medical Company
398th Field Hospital
? Veterinary Hospital
Transport:
? Light Munitions Column

The year 1918 found the division in the lines near the Foret de Holthust. It was relieved on 4 April and sent to Armentières. It followed the German advance and came into action on 11 April north of Armentières. It suffered heavy losses and on 13 April it was withdrawn to rest. On 17 April it went to the Roulers area. On 11 May it was back in the lines north of Hinges. On 25 May it was withdrawn and rested until 11 June, went back to the lines, and was again withdrawn on 22 June. On 26 June it went into the lines in the Artois Region, south of Loos, and held this quiet sector until 2 October. On 5 October it was back in the lines in Belgium, southwest of Roulers, and remained in action

---

121 Disbanded on 8 September 1916.

122 Joined according to *Stein* apparently in August 1916, but was disbanded on 30 December 1916 and distributed among the 8th Dragoons.

123 Raised in July 1916 from the 25th Heavy and the 187th Medium Trench Mortar Detachments, the 299th Light and half of the 187th Medium Trench Mortar Sections. Distributed in autumn 1918 among the division's infantry.

124 On 14 November 1917 joined from the 3rd Guard Cavalry Brigade which had become independent in October 1916. For 1917 the *Histories* also attributes the 1/ and 2/1st Reserve Uhlan Regiment to this division. That has to be an error, as on 30 December 1916 the 1st was disbanded and the 2nd became the 3/4th Reserve Uhlan Schützen Regiment.

125 This battalion is not mentioned by *Stein*. According to him, though, also the 2nd Landsturm Pioneer Company, 5th Corps District, was, for some unspecified time, assigned to the division. It later served in the 231st Pioneer Battalion as army troops.

until 5 November, when it was withdrawn for the last time. In 1918 the division contained:

36th Reserve Division:
  69th Reserve Brigade:
    54th Infantry Regiment
    5th Reserve Regiment
    61st Reserve Regiment
  Cavalry:
    5/2nd Guard Dragoon Regiment
  72nd Artillery Command:
    36th Reserve Field Artillery Regiment (9 btrys)
    3rd Bn/4th Reserve Foot Artillery Regiment
    833rd Light Ammunitions Column
    1252nd Light Ammunitions Column
    1306th Light Ammunitions Column
  2nd Pioneer Battalion:
    1/2nd Pioneer Battalion[126]
    1st Reserve Company, 2nd Pioneer Bn
    236th Trench Mortar Company
    45th Searchlight Section
  436th Signal Command:
    436th Telephone Detachment
    119th Wireless Detachment
  Medical:
    515th Medical Company
    10th Reserve Field Hospital
    12th Reserve Field Hospital
    138th Veterinary Hospital

37th Infantry Division: The 37th Division of the 20th Corps District began the war on the Eastern Front and participated in the Battle of Tannenberg, then marched on Warsaw in October, and fought in the battles on the Rawka during the winter of 1914/1915. In August 1914 the division contained:

37th Infantry Division:
  73rd Infantry Brigade:
    147th Infantry Regiment
    151st Infantry Regiment
    1st Jäger Battalion
  75th Infantry Brigade:
    146th Infantry Regiment
    150th Infantry Regiment
  Cavalry[127]:
    1/ to 4/11th Dragoon Regiment[128]
  37th Field Artillery Brigade:
    73rd Field Artillery Regiment
    82nd Field Artillery Regiment
  Support Troops:
    1/26th Pioneer Battalion
    37th Divisional Bridging Train
    1st Medical Company

In April 1915 the division was on the Narew. In May the 146th Infantry Regiment was sent to the newly forming 101st Infantry Division. The division continued fighting on the Narew until 31 July. By the end of August the division was in Bialystok and 2 September it entered Grodno. It advanced to the Berezina and then in October it

---

126 Joined from the 3rd Infantry Division. According to *Stein*, for some unspecified time, perhaps in 1916/17, also the 279th Pioneer Company served with this division. Before joining, it served with the Searchlight Ersatz Battalion and the Southern Army. It left for the 53rd Reserve Division.

127 According to *Stein* at some time not specified by him the entire 39th Cavalry Brigade of the 4th Cavalry Division was seconded to the 37th Infantry Division including the 14th Dragoon and the 3rd Jäger zu Pferd Regiments. It would, though, soon return to the 4th Cavalry Division. *Voigt* writes that the 3rd Jäger zu Pferd fought in the battles on the Rawka and Bzura Rivers.

128 Mobilized on 2 August 1914 leaving behind the 5th Sqn as its Ersatz Sqn and another part of the Boyen Fortress at Lötzen, East Prussia (it became mobile itself and later the 1/92nd Landwehr Cavalry Regiment; a new Ersatz Sqn was raised in December 1914 at Lyck). On 8 July 1916, the 1st Sqn joined the 11th Landwehr Division, the 3rd the 83rd and the 4th the 89th Infantry Division. The 2nd joined the 101st Infantry Division. A Voigt reports it served in the Balkans from 3 September 1915 and this might have taken place before the regiment was disbanded. It left the 101st for the 11th Army and then consecutively served with the Popadije Detachment, the Bulgarian Nikloff Division, the Dietrich Division and the 9th Austro-Hungarian Division. At the end of the war it was with the 302nd Infantry Division.

was near Dvinsk as the front stabilized. It remained there until it departed for the Western Front in December 1916. In 1915 the division contained:

37th Infantry Division:
  73rd Infantry Brigade:
    147th Infantry Regiment
    151st Infantry Regiment
  75th Infantry Brigade[129]:
    146th Infantry Regiment
    150th Infantry Regiment
  Cavalry:
    11th Dragoon Regiment
37th Field Artillery Brigade:
    73rd Field Artillery Regiment
    82nd Field Artillery Regiment
Engineers:
    1/26th Pioneer Battalion
    37th Divisional Bridging Train
Signals:
    37th Telephone Detachment

In 1916 the 150th Infantry Regiment was temporarily detached to help counter the Russian 1916 offensive on the Stokhod and then, was temporarily assigned to the 91st Division. The division suffered tremendous losses in the battles on the Stokhod, but these were apparently replaced. At the end of September the 150th Infantry Regiment was detached to Galicia, but returned later. On 10 December the division was sent to France. In 1916 the division contained:

37th Infantry Division:
  73rd Infantry Brigade:
    147th Infantry Regiment
    150th Infantry Regiment
    151st Infantry Regiment
  37th Field Artillery Brigade:
    73rd Field Artillery Regiment
    82nd Field Artillery Regiment
Cavalry:
    11th Dragoon Regiment (3 sqns)
Engineers:
    1/26th Pioneer Battalion[130]
    37th Trench Mortar Company[131]
    37th Divisional Bridging Train
Signals:
    37th Telephone Detachment

The division regrouped in Upper Alsace in 1917 and rested until mid-January. It occupied a section of the line from Niederlarg to the Swiss border and remained there until May. About 1 May it was sent to the vicinity of Sissonne. After a short rest it went into the lines on the Aisne at the Chemin des Dames and remained there until the end of July. During this period it was not engaged in any significant action. On 11 July, however, its regiments, supported by the 5th Assault Battalion, reduced the salient near the Cerny sugar refinery. The division lost very heavily in this attack. It was relieved on 31 July and went to the St. Gobain sector, remaining there until the end of November. It suffered heavy losses on 23 October from an Allied bombardment. On 24 October it was withdrawn to a quiet sector. Though recruited in East and West Prussia, the division also received drafts from the Silesian 5th and 6th Corps District. During its sojourn on the Eastern Front it received large numbers of Alsace-Lorrainers. In 1917 the division contained:

---

129 On 15 May 1915 renamed as 201st Infantry Brigade of 101st Infantry Division.

130 According to *Stein* it later changing places with the 3/26th Pioneer Battalion of the 41st Infantry Division.

131 Raised in September 1915 from the 27th Heavy and the 130th Medium Trench Mortar Detachments and the 210th Light Trench Mortar Section. Distributed in autumn 1918 among the division's infantry regiments.

37th Infantry Division:
- 73rd Infantry Brigade:
  - 147th Infantry Regiment
  - 150th Infantry Regiment
  - 151st Infantry Regiment
- Cavalry:
  - 3/10th Jäger zu Pferd Regiment[132]
- 37th Artillery Command:
  - 73rd Field Artillery Regiment
- 134th Pioneer Battalion:
  - 3/26th Pioneer Battalion
  - 250th Pioneer Company
  - 37th Trench Mortar Company
  - 250th Searchlight Section
- Signals:
  - 37th Telephone Detachment
- Medical:
  - 40th Medical Company
  - 317th Field Hospital
  - 318th Field Hospital
  - ? Veterinary Hospital
- Transport:
  - ? Light Munitions Column

During January and February 1918 the division operated in the Soissons Region. It may have been trained in open warfare around this period. On 21 March the division went to Picardy and reinforced the front near Benay. It attacked aggressively. It was withdrawn and rested on 30 March. On 9 April it went into the lines north of Thiescourt and was withdrawn on 20 April. After a period of rest the division attacked, on 27 May, in the Battle of the Aisne, near Presles, south of Laon. It pushed as far as the Troesnes-Longpont sector, when it was relieved on 4 June. On 2 July the division, along with two others, formed the von Conta Group, and on the 15th began to sweep up the Marne Valley. It was hoped that this attack would result in the fall of Rheims and the Montagne de Rheims. This attack, however, was stopped at Bois du Chataignier. It was withdrawn and rested after this attack, taking as replacements men from the disbanded 10th Landwehr Division's 372nd Infantry Regiment. On 13 August the division went into the lines at Verdun and remained there until 20 September. On 26 September it went to the Argonne, near Montfaucon, and launched a counterattack. After being heavily engaged, it was withdrawn. It then moved east in anticipation of an American attack near Exermont. It was engaged in a number of minor actions, which cost it heavily. It is known to have lost 962 men. It was withdrawn and rested on 18 October. On 9 November it went back into the lines and remained there until 11 November. In 1918 the division contained:

37th Infantry Division:
- 73rd Infantry Brigade:
  - 147th Infantry Regiment
  - 150th Infantry Regiment
  - 151st Infantry Regiment
- Cavalry:
  - 3/10th Jäger zu Pferd Regiment
- 37th Artillery Command:
  - 73rd Field Artillery Regiment
  - 2nd, 9th & 10th Btrys, 2nd Bn/16th Foot Artillery Regiment
  - 846th Light Ammunitions Column
  - 924th Light Ammunitions Column
  - 1184th Light Ammunitions Column
- 134th Pioneer Battalion:
  - 3/26th Pioneer Battalion
  - 250th Pioneer Company
  - 37th Trench Mortar Company
  - 63rd Searchlight Section
- 37th Signal Command:
  - 37th Telephone Detachment
  - 82nd Wireless Detachment
- Medical:
  - 40th Medical Company
  - 317th Field Hospital
  - 318th Field Hospital
  - 194th Veterinary Hospital

---

132 Joined in autumn 1916 from the 2nd Infantry Division.

38th Infantry Division: When the war began the 38th Infantry Division from Thuringia, in the 11th Corps DistrictM advanced into Belgium and stood outside Namur until the city surrendered. When East Prussia was invaded by the Russians it went east on 27 August and fought there from 9 to 11 September. It operated in Poland during September and October. In November the division joined Mackensen's 9th Army and in December the 10th Army. It was repeatedly engaged in the winter of 1914/1915 in the battles on the Bzura, the Rawka, and the Pilica. In August 1914 the division contained:

38th Infantry Division:
- 76th Infantry Brigade:
  - 71st Infantry Regiment
  - 95th Infantry Regiment
- 83rd Infantry Brigade:
  - 94th Infantry Regiment
  - 96th Infantry Regiment

Cavalry:
- 2/ to 4/6th Cuirassier Regiment (2nd Half Regiment)[133]

38th Field Artillery Brigade:
- 19th Field Artillery Regiment
- 55th Field Artillery Regiment

Support Troops:
- 2/, 3/11th Pioneer Battalion[134]
- 38th Divisional Bridging Train
- 2nd Medical Company

At the beginning of 1915 the division returned to the 9th Army and was engaged near Rava on 6 and 7 March. It then went north of Warsaw. The 71st Infantry Regiment was detached in May to the new 103rd Division. From July to September the division was repeatedly engaged. It advanced southeast of Bialistok and reached Svislotch on 1 September. On 25 September the division was shipped to the Western Front. It arrived on 1 October and completed its reorganization when its ranks were filled up to replace its losses. It then went into the lines in the Tracy le Val sector in the Oise Region, where it remained quietly until May 1916. In 1915 the division contained:

38th Infantry Division:
- 76th Infantry Brigade[135]:
  - 71st Infantry Regiment
  - 95th Infantry Regiment
- 83rd Infantry Brigade:
  - 94th Infantry Regiment
  - 96th Infantry Regiment
- Attached:
  - 11th Jäger Battalion

Cavalry:
- 3/, 4/6th Cuirassier Regiment (2nd Half Regiment)[136]

38th Field Artillery Brigade:
- 19th Field Artillery Regiment
- 55th Field Artillery Regiment

Engineers:
- 3/11th Pioneer Battalion
- 38th Divisional Bridging Train

Signals:
- 38th Telephone Detachment

---

133 The regiment of six squadrons was mobilized in two half regiments as the divisional cavalry of the 11th Army Corps, the 1st serving with the 22nd and the 2nd with the 38th Infantry Division. In mid-1915 the 1st and 2nd Sqns (under command of Rittmeister von Rochow) joined the 107th Infantry Division, the 3rd and 4th (under command of Rittmeister von Wulffen) remained with the 38th Infantry Division and the Staff, 5th and 6th Sqns with the 22nd Infantry Division. In summer 1916 the squadrons were distributed individually.

134 2/11th Pioneer Battalion joined 22nd Infantry Division.

135 Disbanded on 15 May 1915.

136 By June 1915 the 2nd Sqn joined the 107th Infantry Division and in February 1918 (this could be a printing error in *Stein's* paper; autumn 1916 would be more plausible) the 14th Landwehr Division. Still in February 1918, it formed part of the improvised Rochow Regiment of the Ledebur Cavalry Brigade and in May 1918 joined the 7th Cavalry Schützen Division. The half regiment itself was disbanded on 2 August 1916. While the 3rd Sqn stayed with this division, the 4th went even in July 1916 to the 1st Naval Division and in April 1917 to the 21st Landwehr Division.

On 11 May 1916 the division was sent to the Verdun front. It occupied Hill 301 and remained there until 10 October, when it was so weakened by battle casualties, that it was withdrawn. It had lost 52% of its infantry in the preceding 5 months. On 12 October it was sent to the Somme and went into action on 22 October at Thiepval-Grandcourt, fighting for three weeks and suffering heavy casualties. On 13 November it was withdrawn and rested on the Flanders coast. On 19 December it went back to the Somme. In 1916 the division contained:

38th Infantry Division:
  83rd Infantry Brigade:
    94th Infantry Regiment
    95th Infantry Regiment
    96th Infantry Regiment
  Cavalry:
    3/6th Cuirassier Regiment
  38th Field Artillery Brigade:
    19th Field Artillery Regiment
    55th Field Artillery Regiment
  Engineers:
    3/11th Pioneer Battalion
    38th Trench Mortar Company[137]
    38th Divisional Bridging Train
  Signals:
    38th Telephone Detachment
  Attached:
    25th Labor Battalion

In January 1917 elements of the division were sent north of Courcelette and southwest of Serre as reinforcements to other units. On 17 January it was reunited in the lines near Puisieux-Hébuterne. The division was relieved about 8 March. On 17 March it went into the lines near Beugny Bertincourt, but soon began its retreat on the Hindenburg line under heavy Allied pressure. It took up positions west of Cambrai and remained there until April.

It rested in the Cambrai-Douai area in early May and went into the lines east of Arras on 16 May. It left on 31 May after suffering heavy losses during counterattacks on Rouex. It received 800 men as replacements from the dissolved 624th Infantry Regiment. On 8 June the division was transferred to Flanders, near Gheluwe and finally was sent to the Messines front. On 27 July, before the pending British attack, it went into the lines east of Ypres. When the British finally attacked, on 31 July it suffered heavy losses. The division was relieved on 1 August and sent to Antwerp to rest and reorganize. On 2 September it went into the lines at Monchy le Preux in the Artois, where it suffered heavy losses by artillery fire. On 2 November it was withdrawn and rested near Douai. It returned to the lines north of Ypres from 19 to 25 November, and then went north of Passchendaele, where the British attacked it and inflicted heavy losses on 3 December. It was relieved on 19 November and then rested near Bruges. Although recruited in the small Thuringian states, at the beginning of 1917 a large number of replacements from Baden had been taken in. By the end of that year, though, they had been drafted to Baden regiments in other divisions. In 1917 the division contained:

38th Infantry Division:
  83rd Infantry Brigade:
    94th Infantry Regiment
    95th Infantry Regiment
    96th Infantry Regiment
  Cavalry:
    3/6th Cuirassier Regiment
  38th Artillery Command:
    19th Field Artillery Regiment (9 btrys)

[137] Raised in October 1915 from the 32nd Heavy, the 157th Medium and the 264th Light Trench Mortar Detachments and two thirds of the 216th Light Trench Mortar Section. Distributed in autumn 1918 among the division's infantry regiments.

135th Pioneer Battalion:
  3/11th Pioneer Battalion
  285th Pioneer Company
  38th Trench Mortar Company
Signals:
  38th Telephone Detachment
Medical:
  ? Medical Company
  108th Field Hospital
  ? Veterinary Hospital
Transport:
  ? Light Munitions Column

The division remained in the Passchendaele area until relieved on 5 April 1918. It rested near Lille until 15 April and then went to reinforce the lines at Meteren. It was withdrawn on 8 May and went to Provin where it rested until 21 May. Subsequently it held the sector north of Givenchy until 5 July when it was again relieved and rested at Lille until 6 August. At that time it was alerted and sent to Cambrai. On 10 August it went by motor trucks into the lines near Lihons in the Somme Region. Subsequently the division resisted the advancing Allies. It remained south of Chaulnes until 20 August, when it was driven back to St. Christ. It then continued falling back and was withdrawn from action on 22 September. It rested a week and was then sent to the Cambrai Front, near Rumilly, on 1 October. The division was heavily engaged on 16 October. It had, between 11 August and 1 October, lost over 2,000 prisoners. On 29 October the division was sent to Belgium, and took up a position in the lines near Roubaix until the armistice. In 1918 the division contained:

38th Infantry Division:
  83rd Infantry Brigade:
    94th Infantry Regiment
    95th Infantry Regiment
    96th Infantry Regiment
  Cavalry:
    3/6th Cuirassier Regiment
  38th Artillery Command:
    19th Field Artillery Regiment
    Staff, 1st, 2nd & 3rd Btrys, 61st Foot Artillery Battalion
  135th Pioneer Battalion:
    3/11th Pioneer Battalion
    285th Pioneer Company
    46th Searchlight Section
  38th Signal Command:
    38th Telephone Detachment
    118th Wireless Detachment
  Medical:
    27th Medical Company
    104th Field Hospital
    108th Field Hospital
    38th Veterinary Hospital
  Transport:
    571st Motor Transport Column

38th Landwehr Division: The core of the division, formed in April 1917, was the 38th Mixed Landwehr Brigade. This independent brigade was originally held back in Northern Germany. On 21 October 1914 it arrived at Liege. On 27 October the brigade was sent to Flanders where it held a sector north of the Passchendaele Canal at Nieuport until early November. It then moved to the lines in front of Passchendaele and remained there until the end of December. In 1914 the 38th Landwehr Brigade consisted of:

38th Mixed Landwehr Brigade:
  77th Landwehr Regiment[138]
  78th Landwehr Regiment[139]
  3rd Landwehr Squadron, 10th Corps District[140]
  1st Ersatz Battalion, 59th Field Artillery Regiment

138. Raised in August 1914 in the 10th Corps District by the 78th (at Osnabrück) and 91st Infantry Regiments (3rd Bn; at Oldenburg).

In April 1915 the brigade participated in the Second Battle of Ypres. On 18 May it was transferred to La Bassée. During the summer it rested near Lille and at the end of August went into the lines south of the Lys. Its organization is not known to have changed until 1916.

In 1916 the brigade was withdrawn from the lines near Armentières and sent south of Arras. It continued in the Artois Region until mid-September, when it was sent back to Armentières. It would remain there for the next 18 months. In 1916 the brigade contained:

38th Mixed Landwehr Brigade:
- 77th Landwehr Regiment
- 78th Landwehr Regiment

Cavalry:
- 2nd Landwehr Squadron, 10th Corps District[141] (?)

Artillery:
- 1st Ersatz Battalion, 59th Field Artillery Regiment

When formed in April 1917, the division consisted of the 77th and 78th Landwehr Regiments and the combined 79th Reserve Regiment. In 1917, the division contained:

38th Landwehr Division:
- 38th Mixed Landwehr Brigade:
  - 77th Landwehr Regiment
  - 78th Landwehr Regiment
  - 79th Reserve Regiment[142]
- Cavalry:
  - 3rd Landwehr Squadron, 10th Corps District
- 145th Artillery Command[143]:
  - 255th Landwehr Field Artillery Regiment[144]
- 438th Pioneer Battalion:
  - 5th Landsturm Pioneer Company, 9th Corps District
  - 338th Trench Mortar Company[145]
- Signals:
  - 538th Telephone Detachment
- Medical:
  - 557th Medical Company
  - 109th Reserve Field Hospital
  - 12th Landwehr Field Hospital
  - ? Veterinary Hospital
- Transport:
  - ? Light Munitions Column

---

139 Raised in August 1914 in the 10th Corps District by the 77th (1st, 2nd Bns; at Celle) and 92nd Infantry Regiments (3rd Bn; at Brunswick).

140 Raised in August 1914 by the 17th Hussar Regiment for the 38th Mixed Landwehr Brigade. Passed to the 26th Landwehr Brigade, the Liège Government, then the the 2nd Naval Division and finally returned to the 38th Landwehr Division. Disbanded on 28 September 1918.

141 Raised in August 1914 by the 17th Hussar Regiment for the 37th Mixed Landwehr Brigade. According to *Stein* it passed to the 38th Landwehr Division – this could be a printing error meaning 38th Landwehr Brigade –, the Naval Corps and the 4th Line-of-communication Inspection. Disbanded on 20 February 1917 at home at Brunswick.

142 Raised in August 1914 in the 10th Corps District at Aurich (1st Bn) and Oldenburg, the first two battalions formed the Garrison of Borkum Island while the 3rd Bn and the machine gun company were attached to the 19th Reserve Division. In January 1915 the regimental staff and the 1st Bn (1st, 4th, 5th and 6th Companies; the remaining 2nd, 3rd, 7th and 8th companies formed a new 2nd Bn) were sent to the field and combined in February 1915 with the 1st and 2nd Bn/85th Landwehr Regiment to form the reorganized 79th Reserve Regiment. It joined the 9th Landwehr Division and in April 1917 the 38th Landwehr Division.

The detached 3rd Bn became in August 1916 the 3rd Bn/29th Bavarian Infantry Regiment. On 13 October 1916, it left that regiment again. In November 1916 the 2nd Bn from Borkum was also sent to the field. Subsequently both formed the new 440th Infantry Regiment of the 183rd Division together with the 4th Bn/75th Landwehr Regiment (that Bremen Bn always had been serving detached from its regiment).

In January 1918 the division was withdrawn from the Armentières sector and sent east of Bruges. It rested for a week and went into the lines at Dixmude on 22 January. It was withdrawn on 20 February and on 3 March it relieved the 2nd Naval Division east of Ramscappelle. Around 10 May the division moved slightly south. Then, in mid-October it sidestepped again. However when the armistice came it had been forced to withdraw a considerable distance to the east. Recruited in the 10th Corps District, most men of the division were about 40 years old in 1918. In 1918 the division contained:

38th Landwehr Division:
- 38th Landwehr Infantry Brigade:
  - 425th Infantry Regiment[146]
  - 79th Reserve Regiment
  - 77th Landwehr Regiment
  - 78th Landwehr Regiment
- Cavalry:
  - 2/17th Hussar Regiment[147]
- 145th Artillery Command:
  - 255th Landwehr Field Artillery Regiment
- 438th Pioneer Battalion:
  - 2nd Landsturm Pioneer Company, 2nd Corps Dist
  - 5th Landsturm Pioneer Company, 9th Corps Dist
  - 338th Trench Mortar Company
  - 243rd Searchlight Section
- 538th Signal Command:
  - 538th Telephone Detachment
  - 9th Wireless Detachment
- Medical:
  - 557th Medical Company
  - 109th Reserve Field Hospital
  - 12th Landwehr Field Hospital
  - 538th Veterinary Hospital

39th Infantry Division: When the war began the division from the 15th Corps District defended its home, fighting at the Bonhomme pass in the Alsatian Vosges. On 19 August it was transferred to Dabo and fought at Albreschweiller on the 20th. On 9 September it was sent northwest of Rheims and was engaged between Craonne and Ailles until October. In late October it was sent to Flanders and formed the right wing of the 6th Army, north of Lille. It would remain there until the summer of 1915. In August 1914 the division contained:

39th Infantry Division[148]:
- 61st Infantry Brigade:
  - 126th Infantry Regiment
  - 132nd Infantry Regiment
  - 8th Jäger Battalion
- 82nd Infantry Brigade:
  - 171st Infantry Regiment
  - 172nd Infantry Regiment
  - 14th Jäger Battalion

The 85th Landwehr Regiment had been formed on mobilization in August 1914 in the 9th Corps District by the 86th Fusilier (at Flensburg) and the 84th Infantry Regiments (3rd Bn; at Schleswig) as Garrisons of the Sylt and Pellworm islands. The regimental staff and the 1st and 2nd Bns were sent to the Western Front in autumn 1914, where they were serving in a number of formations. The staff became the regimental staff of the new 362nd Infantry Regiment and the 1st and 2nd Bns became, in autumn 1916, the 2nd and 3rd Bn/79th Reserve Regiment. The 3rd Bn was sent to the Eastern Front in autumn 1914 and became the new 3rd Bn/116th Reserve Regt.

143 Joined on 8 May 1917.

144 Raised in summer 1915 including the 59th Field Artillery Ersatz Battalion.

145 Raised in April 1917. Distributed in autumn 1918 among the division's infantry.

146 Raised on 1 October 1916 by the Supreme Commander East at the Eichhorn Army Group as Field Regiment of the 10th Army with two Prussian and one Saxon battalion. In February 1917 the latter was exchanged for the Prussian 3rd Bn/Saxon 351st Infantry Regiment (originally the Stettin Landsturm Infantry Battalion (18th Landsturm Battalion, 2nd Corps District)), the Saxon battalion taking its place at the Saxon regiment. Replacements were sent from the 1st Ersatz Bn/69th Infantry Regiment (from the 8th Corps District).

147 Would have joined in September 1918, when the 43rd Reserve Division was disbanded.

Cavalry:
1/, 2/, 4/, 5/14th Dragoon Regiment[149]
39th Field Artillery Brigade:
66th Field Artillery Regiment
80th Field Artillery Regiment
Support Troops:
2/, 3/15th Pioneer Battalion
39th Divisional Bridging Train
2nd Medical Company

January 1915 found the division in Flanders south of Ypres on the left wing of the 4th Army. In April the 39th Division surrendered the 171st Infantry Regiment to the newly forming 115th Division. The division remained near Ypres until February 1916. The 172nd Infantry Regiment suffered very heavy losses on 25 September. In January 1915 the division contained:

39th Infantry Division
61st Infantry Brigade:
126th Infantry Regiment
132nd Infantry Regiment
82nd Infantry Brigade[150]:
171st Infantry Regiment
172nd Infantry Regiment
Cavalry:
unknown
39th Field Artillery Brigade:
66th Field Artillery Regiment
80th Field Artillery Regiment
Support Troops:
2/, 3/15th Pioneer Battalion
39th Divisional Bridging Train
Signals:
39th Telephone Detachment

In early 1916 the division was transferred to the Verdun front, concentrating near Piennes-Etain-Ornel-Senon. Early in the 2 February German offensive it suffered few casualties. The division was slowly engaged and on 8 March the 132nd Infantry Regiment took part in the attacks on Fort Douaumont and on the 18th it attacked the Caillette Wood. It suffered enormous casualties. On 11 July the 126th Infantry Regiment was fighting alongside two regiments of the 30th Division. During its actions around Verdun, until 18 August, the division lost 69% of its infantry.

On 20 October the division left Verdun and on 29 October went into the lines on the Somme, in the Sailly Saillisel sector. During a French attack on that sector all three regiments were put into the line simultaneously. They suffered very heavy losses, which reduced the companies to an average of 80 men. The division left the Somme around 11 November and went back to Verdun. From 8 to 12 December it was in the sector between the Louvemont Road and the Chaufour Wood. Here the Allies attacked it on 15 December. Exhausted, the division was relieved on 20 December, rested and reorganized near Vouziers. In 1916 it contained:

39th Infantry Division:
61st Infantry Brigade:
126th Infantry Regiment
132nd Infantry Regiment
172nd Infantry Regiment
39th Field Artillery Brigade:
66th Field Artillery Regiment
80th Field Artillery Regiment
Cavalry:
2/8th Reserve Hussar Regiment[151]

---

148 The 126th Infantry Regiment was a Württemberg regiment and the 14th Jäger a Mecklenburg formation in an otherwise Prussian division. The troops were, respectively, part of the peacetime Garrison of Strassburg Fortress and of Colmar.

149 *Stein* omits the 4th Sqn which most certainly is a printing error. Already on 28 September 1914 the regiment became army cavalry with the 4th Cavalry Division and joined the re-raised 39th Cavalry Brigade of that division.

150 On 2 April 1915 renamed as 229th Infantry Brigade of 115th Infantry Division.

Engineers:
2/, 3/15th Pioneer Battalion
39th Trench Mortar Company[152]
39th Divisional Bridging Train
Signals:
39th Telephone Detachment

Around 10 January 1917 the division went into the lines at Ville sur Tourbe, in the Argonnes. It left in March and was engaged in the Champagne in the attack of 27 March at Cernay en Dormois. It remained in the Massiges sector until early May. From 11 May to early July it was posted near Rheims. After resting near Asfeld, it was sent, in mid-July, to the lines west of Fontaine les Croisilles in the Artois. In September it moved to the Loos sector. In late October the division went to the Passchendaele sector, in Flanders, then to the Becelaere sector. In late November it was sent back to the Artois, north of the La Bassée Canal, where it remained until February 1918. While the 172nd Infantry Regiment received replacements from the Rhine Province, the Rhenish part of the Hessian Province and also from Baden, the Württemberg 126th Infantry Regiment received in 1917 younger personnel from the 123rd, 125th and 126th Landwehr Regiments. In 1917 the division contained:

39th Infantry Division:
61st Infantry Brigade:
126th Infantry Regiment
132nd Infantry Regiment
172nd Infantry Regiment
Cavalry:
5/9th Dragoon Regiment[153]
2/8th Reserve Hussar Regiment
131st Pioneer Battalion[154]:
2/, 3/15th Pioneer Battalion
39th Divisional Bridging Train
39th Artillery Command:
80th Field Artillery Regiment
Signals:
39th Telephone Detachment
Medical:
38th Medical Company
270th Field Hospital
271st Field Hospital
272nd Field Hospital
? Veterinary Hospital
Transport:
? Light Munitions Column

On 25 February 1918 the division began withdrawing from the lines and rested near Lille. During this rest it was trained in open warfare. In mid-March the division moved to the Cambrai Front. Initially in reserve, the division went into action at Beaumetz, west of Cambrai, on 21 March. It was heavily engaged and withdrawn on 23 March. It returned to the lines on 28 March and reached Hébuterne. It was relieved on 6 April and rested near Cambrai.

---

151 When this regiment was disbanded on 17 October 1916, its 1st Sqn joined this division according to *Stein*. Probably in 1918 it was stationed on the Thimougies Training Ground and was disbanded itself on 21 August 1918. The *Histories* attributes both the 1st and 2nd Sqns during 1916 and 1917, but according to *Stein* the 2nd served with the 30th Infantry Division.

152 Raised in October 1915 from the 44th Heavy, the 134th Medium and the 260th Light Trench Mortar Detachments and the one half, each, of the 107th and 207th Light Trench Mortar Sections. Distributed in autumn 1918 among the division's infantry regiments.

153 The regiment was originally army cavalry. According to *Voigt*, it was distributed by squadron as divisional cavalry in February 1917. However, still in spring 1917 the 5th Sqn moved from the 39th to the new 235th Infantry Division. The *Histories* report the squadron back in the 39th Infantry Division after the 235th had been disbanded in August 1918. That is not mentioned by *Stein*, though.

154 Neither *Stein* nor *Cron* mention this battalion. So, chances are that it did not exist.

On 12 April the division moved to the Lys battlefront and on 17 April went into the lines in the Estaires area. The British had held up the German advance here and the division was not immediately needed. On 30 April it went into the line northwest of Merville, but it was not heavily engaged. It was relieved on 26 May and rested near Lille. On 3 July it went into the lines at Vieux Berquin until 14 July. It then rested near Haubourddin and was trained as an assault division.

On 2 August the division began moving into the lines at Neuville-Vitasse south of Arras. In the fighting that followed it lost 1,300 prisoners and many casualties. It was driven back to Cherisy and on 30 August it was withdrawn from the lines and rested. On 18 September the division returned to the lines northwest of Cambrai and was again driven back by heavy Allied attacks. It was relieved on 28 September. It moved to the lines east of Gheluvelt that same day and on 25 October it was withdrawn. The division rested near Audenarde and on 1 November was back in the lines in the Nukerke sector where it appears to have remained until the armistice. In 1918 the division contained:

39th Infantry Division:
- 61st Infantry Brigade:
  - 126th Infantry Regiment
  - 132nd Infantry Regiment
  - 172nd Infantry Regiment
- Cavalry:
  - 2/8th Reserve Hussar Regiment[155]
- 39th Artillery Command:
  - 80th Field Artillery Regiment
  - 406th Foot Artillery Battalion
  - 869th Light Ammunitions Column
  - 1324th Light Ammunitions Column
  - 1325th Light Ammunitions Column
- 136th Pioneer Battalion
  - 2/, 3/15th Pioneer Battalion
  - 39th Trench Mortar Company
  - 58th Searchlight Section
- 39th Signals Command:
  - 39th Telephone Detachment
  - 84th Wireless Detachment
- Medical:
  - 38th Medical Company
  - 270th Field Hospital
  - 272nd Field Hospital
  - 39th Veterinary Hospital
- Transport
  - 572nd Motor Transport Column

40th Infantry Division: The fourth Saxon division of the 19th Corps District, on 13 August it marched into northern Luxembourg and into Belgium on the 18th. It crossed the Meuse above Dinant on the 23rd and entered France. On 30 August it fought at Chesnois, on 1 September at Semide, on 2 September at Somme Py, and on 4 September at Châlons. The division fought in the Battle of the Marne west of Vitry le François. After the battle it retired on Souain. In early October it moved to Lille and when attacked by the British it pulled back to a position between the Ploegsteert Wood and Grenier Wood in Flanders. During this campaign it suffered very heavy casualties, with, as an example, the 9th Company, 104th Infantry Regiment, being reduced to 38 men. In August 1914 the division contained:

- 88th Infantry Brigade:
  - 104th Infantry Regiment
  - 181st Infantry Regiment
- 89th Infantry Brigade:
  - 133rd Infantry Regiment
  - 134th Infantry Regiment
- Cavalry:
  - 1/, 2/, 4/, 5/19th Hussar Regiment[156]
- 40th Field Artillery Brigade:
  - 32nd Field Artillery Regiment
  - 68th Field Artillery Regiment

155 The squadron was disbanded on 21 August 1918. The *Histories* attribute the 5/9th Dragoons to the division during 1918. If that is correct the squadron could only have joined after the 235th Infantry Division had been disbanded in August 1918.

Support Troops:
  2/. 3/22nd Pioneer Battalion
40th Divisional Bridging Train
2nd Medical Company

Through all of 1915 and to August 1916 the division remained in Flanders between the Ploegsteert Wood and Grenier Wood. In March 1915 the 133rd Infantry Regiment was sent to the 24th Division. Elements of the division fought as reinforcements of other formations at Neuve Chapelle in March 1915, at Festubert from May to June, and from June to July in the Artois against the Franco-British offensive. In 1915 the division contained:

40th Infantry Division:
  88th Infantry Brigade:
    104th Infantry Regiment
    181st Infantry Regiment
  89th Infantry Brigade[157]:
    133rd Infantry Regiment
    134th Infantry Regiment
  Attached:
    13th Jäger Battalion
Cavalry:
  2/. 5/19th Hussar Regiment
40th Field Artillery Brigade:
  32nd Field Artillery Regiment
  68th Field Artillery Regiment
Engineers:
  2/. 3/22nd Pioneer Battalion
  40th Divisional Bridging Train
Signals:
  40th Telephone Detachment

On 8 August 1916 the division fought in the Battle of the Somme near Pozières. After a period of violent battles it was totally exhausted and withdrawn. At the end of August it left the Somme for the Neuve Chappel-La Bassée Canal sector, in the Artois, where it remained for six weeks. In mid-October the division returned to the Somme and for three weeks it was heavily engaged. It is known to have lost, from August to October 1916, 6,127 men. About 11 November it was relieved and sent to St. Eloi-Messines. In 1916 the division contained:

40th Infantry Division:
  88th Infantry Brigade:
    104th Infantry Regiment
    134th Infantry Regiment
    181st Infantry Regiment
  Attached:
    13th Jäger Battalion
  40th Field Artillery Brigade:
    32nd Field Artillery Regiment
    68th Field Artillery Regiment
Cavalry:
  2/19th Hussar Regiment
Engineers:
  2/. 3/22nd Pioneer Battalion
  40th Trench Mortar Company[158]
  40th Divisional Bridging Train
Signals:
  40th Telephone Detachment
Attached:
  Anti-Aircraft Detachment

On 26 March 1917 the division left Messines and rested near Renaix. On 23 April it went to Flanders and suffered heavy casualties from the Allied artillery

---

156 Mobilized in August 1914 leaving the 3rd Sqn at home in Grimma, Saxony, as Ersatz Sqn. In 1915 the regiment was split into two half regiments, the 2nd and 5th staying and the 1st and 4th leaving for the 24th Infantry Division. In summer 1916 the regiment was distributed by squadron as divisional cavalry. The 2nd Sqn stayed with the 40th Infantry Division and the 1st with the 24th Infantry Division. The 4th and 5th Sqns joined the 19th Ersatz Division. The 4th. though, passed on to the 219th Infantry Division in spring 1917.

157 Joined on 6 March 1915 the 24th Infantry Division.

158 Raised in December 1915 from the 274th Light and one third of the 182nd Medium Trench Mortar Detachments and the 8th Heavy Trench Mortar Section.

preparations for the Battle of Messines. The 104th Infantry Regiment lost 224 prisoners. Exhausted, the division was withdrawn and rested on the day the battle began. It rested from 7 June to 19 July. On 22 July it went into the lines north of Ypres and on 31 July it was again bombarded. After rest and reorganization the division took over, for several weeks, the Itancourt sector near St. Quentin.. From August to September it received 2,300 replacements from Saxon formations on the Eastern Front, especially from the 244th Reserve Infantry Regiment, the 350th Landwehr Regiment, and from the 19th Landsturm Battalion belonging to the Posen garrison. In addition, Saxon subjects were withdrawn from the 428th Infantry Regiment and the 8th Landsturm Battalion.

On 12 October 1917 the division was sent to Flanders. From 17 to 27 October it occupied the Langewaade-Zevscoten sector, and on 27 October it was heavily attacked yet again, with substantial losses. In early November the division was sent to the Russian Front and took up positions south of Smorgoni until early January 1918. The *Histories* reports that on several occasions, while in France, units of this division deserted their posts and fled behind the lines. In 1917 the division contained:

40th Infantry Division:
- 88th Infantry Brigade:
  - 104th Infantry Regiment
  - 134th Infantry Regiment
  - 181st Infantry Regiment
- Cavalry:
  - 2/19th Hussar Regiment[159]
- 40th Artillery Command:
  - 32nd Field Artillery Regiment
- 141st Pioneer Battalion:
  - 2/, 3/22nd Pioneer Battalion
  - 40th Trench Mortar Company
  - 310th Searchlight Section
- Signals:
  - 40th Telephone Detachment
- Medical:
  - 48th (Saxon) Medical Company
  - 306th Field Hospital
  - 309th Field Hospital
  - ? Veterinary Hospital
- Transport:
  - ? Light Munitions Column

From the beginning of January 1918 until the 15th, the division was seen in the Lake Narotch-Niemen area. Subsequently it was withdrawn and held in reserve near Vilna. By 20 March the division was back in Lorraine. On 15 April it went into the lines near Regnieville in the Woevre Region. The division was withdrawn on 1 June and sent on to the Marne. It went into a quiet sector and remained there until 18 July, when the Allies counterattacked driving the division northward. On 24 July it was withdrawn and rested near Oisy le Verger northwest of Cambrai. On 22 August the division went into the lines near Arras. The same day it launched a counterattack and fought until it was withdrawn on 31 August. The division then rested near Roubaix and on 10 September it went into the lines southeast of Ypres. It lost 1,300 prisoners during these actions. On 8 October it was withdrawn and went to the Courtrai area, where it rested for six days. On 5 October it went into the lines near Gulleghem until about the 26th. On 8 November the division went into the lines near Avelghem and remained there until the armistice was signed. In 1918 the division contained:

---

159 *Stein* reports that, at some time, probably in the early 1917, the 2/7th Hussar Regiment also served with the division coming from the 3rd Infantry Division and leaving for a succession of army corps before settling with the 15th Infantry Division.

40th Infantry Division:
- 88th Infantry Brigade:
  - 104th Infantry Regiment
  - 134th Infantry Regiment
  - 181st Infantry Regiment
- Cavalry:
  - 2/19th Hussar Regiment
- 40th Artillery Command:
  - 32nd Field Artillery Regiment
  - 403rd Foot Artillery Battalion
  - 877th Light Ammunitions Column
  - 960th Light Ammunitions Column
  - 1408th Light Ammunitions Column

141st Pioneer Battalion:
- 2/, 3/22nd Pioneer Battalion[160]
- 40th Trench Mortar Company
- 311th Searchlight Section

40th Signal Command:
- 40th Telephone Detachment
- 171st Wireless Detachment

Medical:
- 48th (Saxon) Medical Company
- 306th Field Hospital
- 309th Field Hospital
- 40th Veterinary Hospital

41st Infantry Division: When the war began the 41st Division of the 20th Corps District stayed on the Eastern Front. It started in East Prussia and went into Poland in October. In early December it was at Lodz and fought on the Rawka in January 1915. In August 1914 the division contained:

41st Infantry Division:
- 72nd Infantry Brigade:
  - 18th Infantry Regiment
  - 59th Infantry Regiment
- 74th Infantry Brigade:
  - 148th Infantry Regiment
  - 152nd Infantry Regiment
- Cavalry:
  - 1/, 2/, 4/, 5/10th Dragoon Regiment[161]

41st Field Artillery Brigade:
- 35th Field Artillery Regiment
- 79th Field Artillery Regiment

Support Troops:
- 2/, 3/26th Pioneer Battalion
- 41st Divisional Bridging Train
- 2nd & 3rd Medical Companies

In February 1915 the division was transferred north of the Vistula and operated between Prasnysz and the Bobr-Narew valley until the summer. In May 1915 the 59th Infantry Regiment was sent to the newly forming 101st Infantry Division. In July the division went to the southern frontier of East Prussia and advanced to Mitau, then onto Jakobstadt. In early 1915 the division contained:

41st Infantry Division:
- 72nd Infantry Brigade[162]:
  - 18th Infantry Regiment
  - 59th Infantry Regiment
- 74th Infantry Brigade:
  - 148th Infantry Regiment
  - 152nd Infantry Regiment
- Cavalry:
  - 10th Dragoon Regiment

41st Field Artillery Brigade:
- 35th Field Artillery Regiment
- 79th Field Artillery Regiment

Support Troops:
- 2/26th Pioneer Battalion[163]
- 41st Divisional Bridging Train
- 41st Telephone Detachment

---

160 2/22nd Pioneer Battalion, according to *Stein*, joined the 24th Infantry Division – apparently in 1917 – and towards the end of the war also the 2nd Reserve Company, 23rd Pioneer Battalion.

161 Mobilized on 2 August 1914, leaving the 3rd Sqn as Ersatz Sqn at Allenstein, East Prussia. On 22 May 1915, the 1st Sqn was sent to the newly formed 101st Infantry Division. It moved on to the German Jäger Division in November 1917. Between May 1915 and July 1916, when the regiment was distributed by squadron, the division retained three squadrons. Subsequently the 2nd Sqn joined the 1st Landwehr Division and was later sent to the 91st Infantry Division while the 5th Sqn was attached to the 109th Infantry Division. The 4th stayed. The *Histories* attributes the 1st to this division in 1917 which would be incorrect.

In 1916 the 10th Dragoon Regiment was probably still present in the 41st Division, but it is uncertain. Until October 1916 the division remained on the Dvina, between Friedrichstadt and Jakobstadt. It was struck by the Russian March offensive and participated in the 10 May German attack. It suffered heavy casualties in both actions. On 21 October the division moved to take part in the Rumanian campaign. It arrived in position on 5 November at Pay, south of Hatszeg in Hungary. It fought near Jiu and advanced almost without fighting after that battle. The division entered Bucharest on 6 December. On 7 December it resumed its pursuit of the Rumanians. Reaching the Sereth River, it encountered the Russians. It remained in line there until 8 February 1917. Its losses were slight in its engagements with the Rumanians and heavier in those with the Russians. In 1916 the division contained:

41st Infantry Division:
- 74th Infantry Brigade:
  - 18th Infantry Regiment
  - 148th Infantry Regiment
  - 152nd Infantry Regiment
- Attached:
  - 9th Landsturm Battalion
- Cavalry:
  - 2/, 4/, 5/10th Dragoon Regiment

41st Field Artillery Brigade:
- 35th Field Artillery Regiment
- 79th Field Artillery Regiment

Engineers & Signals:
- 2/26th Pioneer Battalion
- 41st Trench Mortar Company[164]
- 41st Divisional Bridging Train
- 41st Telephone Detachment

Attached:
- 41st Cyclist Company

Between 8 and 15 February the division was shipped to the Western Front, arriving in Lorraine on 20 February. After a month of rest the division went into line in early May at Bois le Prêtre. During its rest the 148th Infantry Regiment received 600 replacements. Sometime between 6 and 9 May the division was sent to Soissons. After a short stay in the Hurtebise sector, it attacked again on 3 June on the Californie Plateau and suffered serious losses (50-60 men per company in the 152nd Infantry Regiment and heavy losses in the 148th Infantry Regiment). The division remained in the Chevreux sector until 25 June. About 3 July it went to the east of the Butte du Mesnil where it remained, without notable action, until early November. By 12 November the division had been sent to Flanders and alternated in the lines with the 38th Division near Staden and the Houthulst Wood. As the division's home, West Prussia in the 20th Corps District, had few inhabitants, recruits from other districts were taken in, especially from the Upper Silesian 6th Corps District. While it was on the Eastern Front also large numbers of Alsace-Lorrainers were drafted into the division. In 1917 the division contained:

41st Infantry Division:
- 74th Brigade:
  - 18th Infantry Regiment
  - 148th Infantry Regiment
  - 152nd Infantry Regiment

Cavalry:
- 4/10th Dragoon Regiment

41st Artillery Command:
- 79th Field Artillery Regiment

---

162 Joined, on 8 May 1915, the 101st Infantry Division.

163 *Stein* reports that this company for some time served also with the 85th Landwehr Dvision.

164 Raised in May 1916 from the 29th Heavy and the 154th Medium Trench Mortar Detachments and the 271st Light Trench Mortar Section. Distributed in autumn 1918 among the division's infantry regiments.

135th Pioneer Battalion:
- 1/, 2/26th Pioneer Battalion
- 41st Trench Mortar Company

Signals:
- 41st Telephone Detachment

Transport:
- ? Light Munitions Column

Medical:
- Medical Company
- 316th Field Hospital
- 321st Field Hospital
- 323rd Field Hospital
- ? Veterinary Hospital

On 14 January 1918 the division was relieved and rested near Bruges. Subsequently it took over a sector north of Ypres. In late January, though, it was relieved and again rested near Bruges. Its artillery received new guns and it is probably that the division received an intensive course of training during. On 26 February the division was in the lines south of Westroosebeke near Ypres. It was relieved on 4 March and rested in the Turcoing area. On 26 March the division was near Oppy northeast of Arras. It was heavily engaged and its attempted attack broke down after it suffered numerous casualties. Around 30 March the division was withdrawn and in early April it was in the Beaumont-Hamel Region north of Albert.

On 11 June it was relieved and sent to rest and reorganize in the Douai area. On 9 July the division went into the lines east of Amiens at Villers-Bretonneux. Here it was caught in the Allied offensive of 8 August and lost over 1,700 prisoners. It was withdrawn on 10 August and spent two weeks behind the lines. On 25 August the division was back in the lines and heavily engaged, losing more than 800 prisoners and more in killed and wounded. In early September it was relieved and sent to rest and rebuild northeast of Nancy. On 8 September it received men from the disbanded 18th Reserve Regiment of the also disbanded 225th Division. It left Metz on 6 October and reinforced the lines east of Grandpré against the Americans. It was withdrawn on 31 October after suffering heavy losses. On 3 November it was thrown into the line near Nouart in response to the American attack on 1 November. It was withdrawn on the 8th and did not return to the lines before the signing of the armistice. In 1918 the division contained:

41st Infantry Division:
- 74th Infantry Brigade:
  - 18th Infantry Regiment
  - 148th Infantry Regiment
  - 152nd Infantry Regiment
- Cavalry:
  - 4/10th Dragoon Regiment
- 41st Artillery Command:
  - 79th Field Artillery Regiment
  - 5th, 7th & 8th Btrys, 2nd Bn/15th Foot Artillery Regiment
- 26th Pioneer Battalion:
  - 1/, 2/26th Pioneer Battalion
  - 41st Trench Mortar Company
  - 37th Searchlight Section
- 41st Signal Command:
  - 41st Telephone Detachment
  - 96th Wireless Detachment
- Medical:
  - 261st Medical Company
  - 316th Field Hospital
  - 323rd Field Hospital
  - 145th Veterinary Hospital
- Transport:
  - 574th Motor Transport Column
- Attached (elements attached 18 August 1918):
  - 10/7th Reserve Foot Artillery Regiment
  - 41st Foot Artillery Battery
  - 65th Carrier Pigeon Loft
  - 574th Carrier Pigeon Loft
  - 72nd Balloon Squadron
  - 219th Reconnaissance Flight
  - 8th Motor Transport Column
  - 29th Motor Transport Column
  - 245th Motor Transport Column
  - 140th Artillery Observation Section

42nd Infantry Division: In August 1914 the division of the 21st Corps District advanced through French Lorraine and was engaged on 20 August northwest of Dieuze. It attacked Rehainviller and Gerbeviller on the 24th. On 26 August the 131st Infantry Regiment was reduced from 68 officers and 3,017 men to 31 officers and 1,562 men. On 3 September the division was near Moyen-Domptail. From 11 to 13 September it withdrew to Dieuze and on the 18th it moved to Cambrai. On 24 September it was on the Somme, fighting at Gruny and Maucourt. It then took up positions on the Chaulnes Front on the road from Amiens to St. Quentin. By 26 December the 131st Infantry Regiment's losses since August were 87 officers and 3,233 men, over 100%. In August 1914 the division contained:

42nd Infantry Division:
- 59th Infantry Brigade:
  - 97th Infantry Regiment
  - 138th Infantry Regiment
- 65th Infantry Brigade:
  - 17th Infantry Regiment
  - 131st Infantry Regiment
- Cavalry:
  - 1/, 3/, 4/, 5/7th Uhlan Regiment[165]

42nd Field Artillery Brigade:
- 8th Field Artillery Regiment
- 15th Field Artillery Regiment

Support Troops:
- 2/, 3/27th Pioneer Battalion[166]
- 42nd Divisional Bridging Train
- 2nd & 3rd Medical Companies

Through January 1915 the division remained in the lines north of Chaulnes. On 25 January it was moved to the Eastern Front and in early February it was concentrating in East Prussia for Hindenburg's counteroffensive designed to drive the Russians from that province. The division advanced rapidly, reaching Sopockin near Grodno on 20 February. On 9 March it suffered heavy casualties from sharp Russian counterattacks.

165 Mobilized in August 1914 leaving the 2nd Sqn behind as Ersatz Sqn at Saarbrücken/Sarrebruck, 21st Corps District. By September 1914 the regiment was split up:

The 3/ and 4/7th Uhlans were seconded to the 4th Cavalry Division where they formed in November 1914 together with the 4/ and 5/7th Dragoon Regiment (from the 31st Infantry Division) the improvised Flotow Cavalry Regiment. By May 1915 this formation contained just the 4/7th Dragoons and 4/7th Uhlans and was called Flotow Cavalry Detachment and in July 1915 Dieringshofen Cavalry Regiment. The 3/7th Uhlan Regiment together with the 4/1st Jäger zu Pferd Regiment and the 3/1st Uhlan Regiment were combined in the improvised Wedel Cavalry Regiment of the 119th Infantry Division which existed from April 1915 to June 1916.

In December 1914 the staff of the 7th Uhlans became the staff of the improvised Keller Cavalry Brigade of two regiments and the 5th Sqn joined the second of those together with the 1/, 2/ and later also 3/4th Bavarian Light Cavalry Regiment (from the 2nd Bavarian Infantry Division), 4/8th Uhlan Regiment (from the 1st Infantry Division) and 3/6th Dragoon Regiment (from the 25th Infantry Division). *Stein* does not say when those improvised formations were dissolved again. However, the 5th Sqn was divisional cavalry again, on the Eastern Front, by February 1915. So, from autumn 1914 onwards just the 1st Sqn would serve with the division.

Probably in February 1915 the 5th Sqn joined the 31st Infantry Division where it remained until the end of 1917, serving in 1918 on the lines-of-communications behind the Western Front. The 1st Sqn later, perhaps in 1916, passed from the 42nd Infantry Division to the 31st, then on to the independent 9th Landwehr Brigade and to the 94th Infantry Division perhaps in 1917. At the end of that year it was transferred to the west. The 3rd Sqn with a machine gun troop joined the 119th and subsequently the 86th Infantry Division. The 4th joined the 82nd Reserve Division and settled with the 207th Infantry Division – probably in 1917.

166 According to *Stein*, the 2/27th Pioneer Battalion became army troops with the 8th Army and at the end of the war served with the 224th Infantry Division.

On 26 March the division was at Kalwarjia on 2 April near Mariampol. It fought there from 29 March to 21 April, then remained there until August. On 30 August it arrived in Vilna. It continued pushing east and when the lines stabilized it was near Postavy. In autumn the 97th Infantry Regiment was transferred to the newly forming 108th Division. From January to August 1915 the division contained:

42nd Infantry Division:
- 59th Infantry Brigade[167]:
  - 97th Infantry Regiment
  - 138th Infantry Regiment
- 65th Infantry Brigade:
  - 17th Infantry Regiment
  - 131st Infantry Regiment
- Cavalry:
  - 1/7th Uhlan Regiment

42nd Field Artillery Brigade:
- 8th Field Artillery Regiment
- 15th Field Artillery Regiment

Engineers:
- 3/27th Pioneer Battalion
- 42nd Divisional Bridging Train

Signals:
- 42nd Telephone Detachment

The division remained in its positions on Lake Narotch during 1916 and until April 1917. However, in late March it was heavily attacked by the Russians and again suffered heavy losses. In 1916 the division contained:

42nd Infantry Division:
- 65th Infantry Brigade:
  - 17th Infantry Regiment
  - 131st Infantry Regiment
  - 138th Infantry Regiment
- 42nd Field Artillery Brigade:
  - 8th Field Artillery Regiment
  - 15th Field Artillery Regiment

Cavalry:
- 1/7th Uhlan Regiment[168]
- 1/, 2/7th Dragoon Regiment[169]

Engineers:
- 3/, 5/27th Pioneer Battalion
- 42nd Trench Mortar Company[170]
- 42nd Divisional Bridging Train

Signals:
- 42nd Telephone Detachment

On 24 April 1917 the division was shipped to the Western Front, but en route it was redirected and sent to Lemberg. It served as a reserve at first, but on 20 July it was part of the German counteroffensive at Brzezany. On 31 July the division had arrived near Tarnopol. It was withdrawn from the Galician front in early August, then moved to Neugut in Courland and took part in the advance on Riga. In late September it moved to Libau and began preparations for the occupation of the Baltic islands. The 131st Infantry Regiment landed on Oesel Island on 12 October and remained there until 1 November. The 138th Infantry remained at Moon until 25 October. In early November the division moved to the vicinity of Kovel. At the end of November the division took over a calm sector in the vicinity of Kachovka. On 23 December it began its movement to France, arriving there on 28 December. As Alsace-Lorrainers were not seen to be dependable, the division recruited by 1917 predominantly in Westphalia and the Rhine Province – 7th and 8th Corps Districts. However, on the Russian Front there were many Alsace-Lorrainers in the division. In 1917 the division contained:

---

167 Disbanded on 7 November 1915.

168 Probably in the second half of 1916 the 1st Sqn joined the 31st Infantry Division.

169 Joined in the second half of 1916, the 2nd Sqn passing on to the 17th Landwehr Division probably in October or November 1916.

170 Raised in October 1916 (?). Distributed in autumn 1918 among the division's infantry regiments.

42nd Infantry Division:
  65th Infantry Brigade:
    17th Infantry Regiment
    131st Infantry Regiment
    138th Infantry Regiment
  Attached:
    Guard Reserve Jäger Battalion[171]
  Cavalry:
    1/7th Dragoon Regiment
    3/, 4/2nd Leibhusaren Regiment[172]
    5/8th Uhlan Regiment[173]
  42nd Artillery Command:
    15th Field Artillery Regiment
  1st Field Bn/27th Pioneer Battalion or 136th Pioneer Battalion:
    2/, 5/27th Pioneer Battalion
    42nd Trench Mortar Company
    229th Searchlight Section
    345th Searchlight Section
  Signals:
    42nd Telephone Detachment
  Medical:
    263rd Medical Company
    ? Field Hospital
    ? Veterinary Hospital
  Transport:
    ? Motor Transport Column

After resting near Lille the division moved east of Armentières on 23 January 1918. On 22 March 1918 it was relieved from its Armentières sector. On 9 April it reinforced the battle front near Merris. Heavily engaged between 9 and 16 April it suffered 50% casualties. It was relieved on 17 April. Then the division went to the quiet Lens sector and remained there until 26 June. On 25 June it moved southwest of Soissons losing 1,400 prisoners in a French attack on 18 July. Withdrawn on 22 July the division was rebuilt near Laon from the recently dissolved 211th Division. After this the division moved to Rethel. On 22 August it entered the Champagne lines and until 1 October it was engaged by the French. During this period it lost 2,000 prisoners. It was withdrawn on 1 October. On 14 October it was back in the lines near Olizy and remained there until the armistice. In 1918 the division contained:

42nd Infantry Division:
  65th Infantry Brigade:
    17th Infantry Regiment
    131st Infantry Regiment
    138th Infantry Regiment
  Attached:
    Guard Reserve Jäger Battalion
  Cavalry:
    1/7th Dragoon Regiment
  42nd Artillery Command:
    15th Field Artillery Regiment
    11th & 12th Btrys, 2nd Bn/15th Foot Artillery Regiment
  27th Pioneer Battalion:
    2/, 5/27th Pioneer Battalion
    42nd Trench Mortar Company
    14th Searchlight Section
  42nd Signal Command:
    42nd Telephone Detachment
    147th Wireless Detachment
  Transport:
    575th Motor Transport Column

---

171 Raised in August 1914 on mobilization at Potsdam by the Guard Jäger Battalion. As the Guard Jäger Battalion tried to keep up the tradition by only recruiting professional hunters and foresters, this would also apply to the reservists of which this battalion was made up.

172 *Stein* reports that for some time also the 3/ and 4/2nd Leibhusaren Regiment (the second of the two "Death Head Hussars") were attached to the reinforced division. This would be, according to him, the last posting of the 4th Sqn while the 3rd left for the North Corps and the 23rd Cavalry Brigade. *Voigt*, however, reports that both squadrons were on the Baltic Islands and afterwards at Orsha. Thus, the posting of the two squadrons to the 42nd Infantry Division was in autumn 1917 as they were still in September 1917 with the reinforced Leibhusaren Brigade, as the division left for the Western Front in December 1917 while the hussars stayed east and as he reports the 1/, 2/ and 4/2nd Leibhusaren Regiment in Finland in 1918.

173 According to *Stein* it joined the 6th Reserve Division and left again for the 221st Infantry Division. So, it might have served at some time during 1917 in this division.

Medical:
263rd Medical Company
269th Field Hospital
368th Field Hospital
162nd Veterinary Hospital

43rd Reserve Division: This division was the first of the second wave of reserve divisions (43rd to 54th and Bavarian 6th Reserve Divisions) raised in September and October 1914, mostly from young wartime volunteers with no prior military experience and who joined directly from school or university. Those "young divisions" could be compared to the British "Kitchener's New Army". As they were engaged prematurely without sufficient training and staffed by inexperienced cadres, so they incurred horrendous casualties in their first battles, especially in promising officer and NCO "material". These missing potential officers and NCOs would be sorely missed in the subsequent years of the war. Learning the hard way, starting with the third wave of winter 1915/1915 (75th to 82nd and Bavarian 8th Reserve Divisions), the German Armies would, in future, form new divisions behind the lines, in the line-of-communications areas. Existing regiments would provide entire companies etc., as cadres.[174] Even if raised at home, raw recruits were mixed with battle experienced cadres recalled from the front and with reconvalescents returning to the lines.

In particular, this division was formed from the regimental recruit depots of the Guard and, as a result, drew recruits from all across the Prussian monarchy. It would continue to do so as the war went on. The division was trained during September on the Doeberitz Training Ground near Berlin and did only entrain for Belgium on 13 October. On 19 October it was engaged near Dixmude, Mercken, and Bixschoote and it remained in action until the end of November. After the Battle of the Yser elements of the division were spread along the front between Ypres and Nieuport. In October 1914 the division contained:

43rd Reserve Division:
85th Reserve Brigade:
1st/, 2nd/, 3 rd Bn/201st Reserve Regiment[175]
1st/, 2nd/, 3 rd Bn/202nd Reserve Regiment[176]
86th Reserve Brigade:
1st/, 2nd/, 3 rd Bn/203rd Reserve Regiment[177]
1st/, 2nd/, 3 rd Bn/204th Reserve Regiment[178]
Attached:
15th Reserve Jäger Battalion[179]
Cavalry:
43rd Reserve Cavalry Detachment[180]
Artillery:
1st/, 2nd/, 3 rd Bn/343rd Reserve Field Artillery Regiment (9 btrys)[181]

174 If individuals had to be seconded, every unit will try to keep the best and transfer non-performers.

175 Raised at Berlin by the Ersatz Battalions of the 2nd and 4th Foot Guards and the Guard Fusilier Regiment. Disbanded on 16 September 1918 and distributed among the 1st and 4th Guard Divisions. As with all of the "young regiments", the 201st was raised with twelve rifle companies, formed in three battalions, and one machine gun section. This would soon be expanded into a machine gun company. The subsequent development of the regimental organization would follow the general pattern.

176 Raised at Berlin by the Ersatz Battalions of the 1st, 2nd and 4th Grenadier Guards. Only 20% of the personnel had already served in peace! Disbanded on 16 September 1918 and distributed among the 2nd Guard and 79th Reserve Divisions.

177 Raised at Spandau and Charlottenburg by the Ersatz Battalions of the 5th Foot Guards, 5th and 3rd Grenadier Guards. Disbanded in September 1918 and distributed among the Guard Ersatz and 79th Reserve Divisions.

178 Raised by the Ersatz Battalions of the 1st and 3rd Foot Guards and the Infantry Training Regiment.

In early January 1915 the 86th Reserve Division was in the lines at Westende. The rest of the division continued operating along the front in Belgium. By late February the division was re-assembled and rested near Menin-Roulers until 25 April. In May elements of the division continued in the Bixschoote-Boesinghe sector, while the rest of the division went to Arras to oppose the French offensive. The 202nd Reserve Infantry Regiment lost 76 officers and 1,320 men at Notre Dame de Lorette.

In early July 1915 the 86th Reserve Brigade was transferred to Russia and participated in Mackensen's offensive into Poland. Between 15 May and 29 September the 204th Reserve Regiment lost 63 officers and 3,511 men. The 85th Reserve Brigade went first to Lorraine, then from July to September to the Woevre Region and took part in the Battle of Champagne in late September. It then joined the rest of the division in Serbia where that took part in the October campaign. In 1915 the division contained:

43rd Reserve Division:
- 85th Reserve Brigade:
  - 201st Reserve Regiment
  - 202nd Reserve Regiment
- 86th Reserve Brigade:
  - 203rd Reserve Regiment
  - 204th Reserve Regiment
- Attached:
  - 15th Reserve Jäger Battalion

Cavalry:
- 43rd Reserve Cavalry Detachment

Artillery:
- 43rd Reserve Field Artillery Regiment (9 btrys)

Engineers:
- 43rd Reserve Pioneer Company
- 43rd Reserve Divisional Bridging Train

Medical:
- 43rd Reserve Medical Company

In late January 1916 the division left Serbia for France, arrived in early February and began rebuilding. After resting near Valenciennes, in late March it went to Verdun and went into action on 10 April, attacking at Bethincourt and Mort Homme. It suffered heavy losses between 10 April and 25 May. In late May the division was withdrawn and rebuilt again, having lost 50% of its infantry.

In mid-June the division once again moved to Russia. It was engaged west of Loutsk in the German June counteroffensive. Its losses were heavy. On 15 November the division went back to France. During this process the 204th Reserve Infantry Regiment was detached and assigned to the new 218th Division. The division rested for a month near Rethel, then moved to Verdun after the French attack of 15 December. On 17 December it went into the lines in the Chambrettes sector. In 1916 the division contained:

43rd Reserve Division:
- 85th Reserve Brigade:
  - 201st Reserve Regiment
  - 202nd Reserve Regiment
- 86th Reserve Brigade:
  - 203rd Reserve Regiment
  - 204th Reserve Regiment
- Attached:
  - 15th Reserve Jäger Battalion

---

179 Raised in autumn 1914 at Potsdam by the Ersatz Detachment/Guard Jäger Battalion. About 80 to 100 per company had already served in peace. The remainder was wartime volunteers. Joined in May 1918 the 14th Jäger Regiment.

180 Raised on 27 August 1914 by the Guard Corps District.

181 Before the war, the virtues and vices of 6-gun and 4-gun batteries had been vehemently discussed. The new field artillery regiments starting with the second wave of reserve divisions received 4-gun batteries because of the sheer lack of guns. Subsequently, all other field batteries were reorganized in the same manner.

Cavalry:
  43rd Reserve Cavalry Detachment[182]
  2/17th Hussar Regiment[183]
Artillery:
  43rd Reserve Field Artillery Regiment

Engineers:
  43rd Reserve Pioneer Company
  43rd Reserve Divisional Bridging Train
  243rd Trench Mortar Company[184]

Despite the heavy losses in late 1916 and early 1917, the division was not seriously engaged prior to its departure from the Verdun Region on 31 January. Nevertheless losses were high. After resting in Alsace the division went to Champagne and on 22 February it went into the lines. A French attack on 16 April inflicted heavy losses. The division was relieved in late April and reorganized. Among others part of the dissolved 613th and 614th Infantry Regiments were drafted. There were also 3,000 replacements from depots in the Brandenburg Province. On 9 May the division went to the Vauquois sector in the Argonnes. In late May the division rested and rebuilt in the Ardennes and then moved and continued the process near Laon. From 18 to 20 July the division was in the line south of Pargny-Filain in the Chemin des Dames Region and was subjected to artillery preparation and attack by the French on 30 July. It suffered heavy losses, which were increased in its counterattacks attempted on 31 July and 10 August. On 10 August the 201st Reserve Infantry Regiment was almost completely destroyed, and lost 124 men as prisoners. The division was relieved from the Chemin des Dames on 23 August and rested until late September near Laon, where it was rebuilt and reorganized. In early October it was trained for an anticipated offensive. It was engaged on 15 October and was attacked on 23 October, losing 53 officers and 2,190 men as prisoners and suffering numerous casualties. The division was relieved on 28 October and sent to the Russian Front, where it arrived on 11 November, near Baranovitchi. By 1917 trained reconvalescents from the other Guard and Guard reserve divisions, as well as from Guard Landsturm battalions were taken in. In 1917 the division contained:

43rd Reserve Division:
  85th Reserve Brigade:
    201st Reserve Regiment
    202nd Reserve Regiment
    203rd Reserve Regiment
  Cavalry:
    2/17th Hussar Regiment
  Attached:
    15th Reserve Jäger Battalion
  43rd Artillery Commend:
    43rd Reserve Field Artillery Regiment

343rd Pioneer Battalion:
  43rd Reserve Pioneer Company
  4/17th Pioneer Battalion
  243rd Trench Mortar Company
Signals & Transport:
  443rd Telephone Detachment
  730th Motor Transport Column
Medical:
  525th Medical Company
  72nd Reserve Field Hospital
  73rd Reserve Field Hospital
  ? Veterinary Hospital

The division left Russia on 9 February 1918 and arrived near Magdeburg in mid-February. It rested five weeks and on 18 March began moving to Loos. The division was engaged on 9 April at Festubert. Its attack there was blocked by the British. On 29

---

182 According to *Stein*, at some unspecified time, this detachment joined the 12th Landwehr Division and later the 2nd Austro-Hungarian Army. Leaving the division would have happened in summer 1916.

183 According to *Voigt*, joined on 18 September 1916.

184 Raised in 1916 from the 112th Medium and half of the 121st Light Trench Mortar Detachments and the 46th Heavy Trench Mortar Section. Disbanded in September 1918.

April it was again relieved. The division rested south of Lille until 24 June and then went into the lines at Bouzencourt. Until 8 August it held that sector on the Somme. On 20 August it went to the lines near Bray and was engaged until late August. It lost a total of 1,100 prisoners in its two engagements in August. The division was disbanded in September 1918. The 203rd Reserve Regiment went to the Guard Ersatz Division and the 202nd Reserve Regiment went to the 2nd Guard Division. Prior to dissolution, the division contained:

43rd Reserve Division:
- 85th Reserve Brigade:
  - 201st Reserve Regiment
  - 202nd Reserve Regiment
  - 203rd Reserve Regiment
- Attached:
  - 15th Reserve Jäger Battalion
- Cavalry:
  - 2/17th Hussar Regiment[185]
- 43rd Artillery Commend:
  - 43rd Reserve Field Artillery Regiment
- 343rd Pioneer Battalion[186]:
  - 4/17th Pioneer Battalion[187]
  - 1st Reserve Company, 23rd Pioneer Battalion[188]
  - 43rd Reserve Pioneer Company[189]
  - 243rd Trench Mortar Company
- Signals:
  - 443rd Telephone Detachment
- Medical:
  - 525th Medical Company
  - 72nd Reserve Field Hospital
  - 237th Veterinary Hospital
- Transport
  - 730th Motor Transport Column

44th Reserve Division: The 44th Reserve Division was formed in September and October 1914 as part of the second wave of reserve divisions in the Brandenburg March, 3rd Corps District, and trained at the Jüterbog Training Ground south of Berlin. One reserve regiment was seconded from the 10th Corps District. It went to action at Dixmude and Bixschoote in October and November and suffered heavy losses. On 9 November the 3rd Bn/205th Reserve Infantry Regiment, was reduced from to 1,044 officers and men to 153! After the Battle of the Yser it occupied several sectors north of Ypres. In October 1914 the division contained:

44th Reserve Division:
- 87th Reserve Brigade:
  - 1st/, 2nd/, 3 rd Bn/205th Reserve Regiment[190]
  - 1st/, 2nd/, 3 rd Bn/206th Reserve Regiment[191]
- 88th Reserve Brigade:
  - 1st/, 2nd/, 3 rd Bn/207th Reserve Regtt[192]
  - 1st/, 2nd/, 3 rd Bn/208th Reserve Regt[193]
- Attached:
  - 16th Reserve Jäger Battalion[194]
- Cavalry:
  - 44th Reserve Cavalry Detachment[195]
- Artillery:
  - 1st/, 2nd/, 3 rd Bn/44th Reserve Field Artillery Regiment (9 btrys)
- Engineers:
  - 44th Pioneer Company

185 Joining the 38th Landwehr Division.

186 Joining the Boehm Army Group according to *Stein*.

187 Joining the 346th Pioneer Battalion according to *Stein*.

188 Joining the 5th Army and then the 218th Pioneer Battalion, according to *Stein*.

189 Joining the 49th Reserve Division.

190 Raised in the 3rd Corps Districtm at Frankfurt/Oderm by the Ersatz battalions of the 8th Grenadier, 48th Infantry and 12th Grenadier Regiments. Per company about 70 volunteers. Remainder reservists and Landwehr men of Berlin so that about 2/3 had already served in peace.

191 Raised in the 3rd Corps District at Brandenburg on the Havel (the town which gave the March or Province of Brandenburg its name) by the Ersatz Battalions of the 35th Fusilier, 20th and 52nd Infantry Regiments. Only about 25 to 30% having served in peace.

The division remained in Flanders until 7 June when the division was transferred to the Eastern Front. It arrived in mid-June and participated in Mackensen's offensive, from 22 June to 16 July in the Battles of Pursuit on the Galician frontier; 19 to 28 July in the Battle of Krasnostaw, on 29 and 30 July in the Battle of Biskupice, from 31 July to 19 August in the Battles of the Bug and on 26 August in the capture of Brest-Litovsk. In October it was sent to Serbia and fought there through the entire campaign. In late December it was sent to rest in Hungary. In 1915 the division contained:

44th Reserve Division:
- 87th Reserve Brigade:
  - 205th Reserve Regiment
  - 206th Reserve Regiment
- 88th Reserve Brigade:
  - 207th Reserve Regiment
  - 208th Reserve Regiment
- Attached:
  - 16th Reserve Jäger Battalion

Cavalry:
- 44th Reserve Cavalry Detachment

Artillery:
- 44th Reserve Field Artillery Regiment

Engineers:
- 44th Reserve Pioneer Company
- 44th Reserve Divisional Bridging Train

Medical:
- 44th Reserve Medical Company

In late January 1916 the division moved to France and arrived on 6 February. It spent some time on the Somme Front and then moved to Landrecies on 24 March. In late March the division concentrated near Buzancy and went to the left bank of the Meuse. On 11 April the 88th Reserve Brigade went into the line in the Mort-Homme sector of the Verdun Region. The whole division was in action on 25 April and suffered very heavy losses from April to May. It was withdrawn from the front on 6 June, rested and reorganized near Sedan and received replacements from its home 3rd and the Silesian 5th Corps District. On 3 July it was transferred to the Somme and some of its elements went into action on the Estrees-Belloy Front on 4 July. After suffering French attacks from 6 to 10 July it counterattacked on the 7th and 8th. In addition to severe casualties, it lost 9 officers and 522 men as prisoners. It was relieved and rested from 10 to 20 July, then went into the lines in the Lassigny-Beuvraignes sector. During this period the division received a large quantity of replacements. In mid-September it went into action on the Somme and between 9 to 28 October it lost heavily. The division then returned to the Lassigny sector where it was reorganized and received 300 to 400 replacements per regiment. At this time it transferred the 207th Reserve Infantry Regiment to the newly forming 228th Division. In 1916 the division contained:

44th Reserve Division:
- 87th Reserve Brigade:
  - 205th Reserve Regiment
  - 206th Reserve Regiment
- 88th Reserve Brigade[196]:
  - 207th Reserve Regiment
  - 208th Reserve Regiment

---

192 Raised in the 3rd Corps District at Prenzlau by the Ersatz Battalions of the 24th (3rd Bn) and 64th Infantry Regiments.

193 Raised in the 10th Corps District by the Ersatz Battalions of the 79th (1st Bn; at Brunswick), 77th (2nd Bn; at Celle) and 79th Infantry Regiments (3rd Bn; at Hildesheim).

194 Raised in autumn 1914 at Lichterfelde near Berlin by the Ersatz Detachment/Guard Schützen Battalion. Joined in August 1916 the 5th Jäger Regiment.

195 Raised on 27 August 1914 by the Guard Corps District. Probably late in the war at some unspecified time it joined, according to *Stein*, the 8th Army.

196 Disbanded on 10 January 1917.

Attached:
- 16th Reserve Jäger Battalion

Cavalry:
- 44th Reserve Cavalry Detachment

Artillery:
- 44th Reserve Field Artillery Regiment

Engineers:
- 5/29th Pioneer Battalion
- 44th Reserve Pioneer Company
- 244th Reserve Trench Mortar Company[197]
- 44th Reserve Divisional Bridging Train

In March 1917 the division took part in the German retreat and took up a position between La Fère and Moy on 25 March. At the end of March the division went into reserve near Marle St. Gobert and on 20 April it concentrated near Monampteuil-Filain. On 21 April it went into position on both banks of the Oise-Aisne Canal. During the night of 5/6 May it was relieved, very exhausted and had lost 1,670 prisoners; it was transferred to the Woevre Region and then to the Côtes de Meuse, northeast of St. Mihiel. It was withdrawn on 25 October and sent to Flanders, north of Passchendaele, on 10 November. In late November it was sent to the Neuve Chapelle sector in the Artois, where it remained until 19 March 1918. By 1917 replacements, practically without exception, came from the Brandenburg 3rd Corps District. In 1917 the division contained:

44th Reserve Division:
- 87th Reserve Brigade:
  - 205th Reserve Regiment
  - 206th Reserve Regiment
  - 208th Reserve Regiment
- Cavalry:
  - 4/7th Dragoon Regiment[198]
- 44th Artillery Command:
  - 44th Reserve Field Artillery Regiment
- 344th Pioneer Battalion:
  - 5/29th Pioneer Battalion
  - 44th Reserve Pioneer Company
  - 244th Trench Mortar Company
  - 306th Searchlight Section
- Signals:
  - 444th Telephone Detachment
- Medical & Transport:
  - 526th Medical Company
  - 73rd Reserve Field Hospital
  - 238th Veterinary Hospital
  - ? Motor Transport Column

Around 1 April 1918 the division withdrew to the second line and returned to the front lines to launch an attack on 13 April near Locon near the La Bassée Canal. It remained in that Region until relieved on 6/7 May. It rested at Courrières for three weeks and then went into the lines west of Vieux Berqum. It remained there until 4 July. The division went northwest of Tournai shortly later, rested and received drafts. It returned to the lines northeast of Martinpuich on 26 August. On the 27th the division fell back to Flers and continued withdrawing until it was southwest of Havrincourt on 7 September. It was withdrawn from the lines on 10 September, after having suffered 700 prisoners and many casualties. The division rested for four weeks and then appeared in the lines on the Cambrai-St. Quentin Front on 10 October, north of Montay. It fought around Le Cateau until the end of the month and then was withdrawn. On 4 November it was back in the lines at Locquignol and fell back towards Meubeuge as the war ended. In 1918 the division contained:

---

197 Raised in December 1915 from the 38th Heavy, two thirds of the 121st Medium and the 221st, 222nd and 229th Light Trench Mortar Detachments.

198 Probably joined in 1917, replacing the 44th Cavalry Detachment before leaving, according to *Stein*, for the 201st Infantry Division. The *Histories* attributes this squadron to both divisions during 1918!

44th Reserve Division:
- 87th Reserve Brigade:
  - 205th Reserve Regiment
  - 206th Reserve Regiment
  - 208th Reserve Regiment
- Cavalry:
  - unknown
- 44th Artillery Command:
  - 44th Reserve Field Artillery Regiment
  - 4th & 6th Btrys, 2nd Bn/21st Foot Artillery Regiment
  - 706th Light Ammunitions Column
  - 828th Light Ammunitions Column
  - 1322nd Light Ammunitions Column
- 344th Pioneer Battalion:
  - 5/29th Pioneer Battalion
  - 44th Reserve Pioneer Company
  - 244th Trench Mortar Company
  - 21st Searchlight Section
  - 250th Searchlight Section
- 444th Signal Command:
  - 444th Telephone Detachment
  - 86th Wireless Detachment
- Medical:
  - 526th Medical Company
  - 73rd Reserve Field Hospital
  - 75th Reserve Field Hospital
  - 238th Veterinary Hospital
- Transport:
  - 731st Motor Transport Column

44th Landwehr Division: The nucleus of the 44th Landwehr Division was the independent Rosenberg Brigade, which was formed in December 1915 with the 1st Landwehr Ersatz Regiment and the 93rd Landwehr Regiment. In July 1916, it was renamed the 44th Landwehr Brigade. Subsequently, it contained:

Rosenberg (44th) Landwehr Brigade:
- 93rd Landwehr Regiment[199]
- 382nd Landwehr Regiment[200]

The division was raised in April 1917 by adding the 81st Landwehr Regiment to this 44th Landwehr Brigade. The latter regiment had fought at St. Dié in spring 1916, and subsequently in the Verdun Region until the end of the year. The 44th Landwehr Brigade was on the left bank of the Moselle and the division after being formed stayed there in the Bois le Prêtre until October 1917. It was relatively inactive. On 13 October it was relieved and sent to the Upper Alsace. On 19 October the division was in the lines on the Rhône-Rhine Canal. It consisted of quite old personnel, however there was an assault party in every regiment as customary. When organized, the division contained:

44th Landwehr Division:
- 44th Landwehr Brigade:
  - 81st Landwehr Regiment[201]
  - 93rd Landwehr Regiment
  - 382nd Landwehr Regiment
- Artillery:
  - 270th Field Artillery Regiment

---

199 Formerly the von Gundlach Regiment of the Graudenz Corps. It had been formed in 1914 from the 3rd Bernburg and 6th Altenburg Landwehr Battalions, 4th Corps District (of the six Landwehr battalions "beyond the establishment", that is, exceeding the mobilization plan of the 4th Corps District) and of the 38th Landwehr Brigade Ersatz Battalion, 10th Corps District, coming from the 3rd Landwehr Ersatz Regiment. The 93rd Landwehr Regiment had been attached to the 8th Ersatz Division at the beginning of 1916.

200 The 1st Landwehr Ersatz Regiment was raised in October/November 1914 with three battalions of the 7th Corps District (25th, 26th, 27th Landwehr Brigade Ersatz Battalions) and one of the 9th (33rd Landwehr Brigade Ersatz Battalion). At the end of 1915, the 7th Corps District's Landwehr Brigade Ersatz Battalions formed the new 382nd Landwehr Regiment, while the regimental staff and the 33rd Landwehr Brigade Ersatz Battalion, to which were added the 9th Landwehr Brigade Ersatz Battalion of the 4th Corps District and the 5th Metz Landsturm Infantry Battalion (3rd Landsturm Battalion, 16th Corps District), formed the new 86th Landwehr Regiment.

Cavalry:
4/2nd Uhlan Regiment[202]
444th Pioneer Battalion:
411th Trench Mortar Company[203]
Signals:
411th Telephone Detachment

Medical:
274th Medical Company
360th Field Hospital
? Veterinary Hospital
Transport:
? Motor Transport Column

In 1918 the division remained in the Altkirch sector until 11 November. This was an absolutely quiet sector. In 1918 the division contained:

44th Landwehr Division
44th Landwehr Brigade:
81st Landwehr Regiment
93rd Landwehr Regiment
382nd Landwehr Regiment
Cavalry:
4/2nd Uhlan Regiment
154th Artillery Command[204]:
4/, 5/, 6/, 8/254th Landwehr Field Artillery Regiment
828th Field Artillery Battery
837th Field Artillery Battery

444th Pioneer Battalion:
2nd Landsturm Pioneer Company, 7th Corps District
4th Landsturm Pioneer Company, 7th Corps District
411th Trench Mortar Company
Signals & Transport:
411th Telephone Detachment
? Motor Transport Column
Medical:
274th Medical Company
268th Field Hospital
360th Field Hospital
? Veterinary Hospital

45th Reserve Division: The 45th Reserve Division was formed in September and October 1914 as a division of the second wave of reserve divisions. It was recruited in the 2nd and 9th Corps Districts and trained at Jüterbog. By 21 October it was in Belgium and engaged in the Battle of the Yser. Between 15 October and 11 November the 212th Reserve Infantry Regiment suffered 82 officers and 1,669 men, which is an indication of the level of overall losses it suffered at this time. In December the division was in the vicinity of Bixschoote. When formed the division contained:

89th Reserve Brigade:
1st/, 2nd/, 3rd Bn/209th Reserve Regiment[205]
1st/, 2nd/, 3rd Bn/212th Reserve Regiment[206]
90th Reserve Brigade:
1st/, 2nd/, 3rd Bn/210th Reserve Regiment[207]
1st/, 2nd/, 3rd Bn/211th Reserve Regiment[208]

Attached:
17th Reserve Jäger Battalion[209]
Cavalry:
45th Reserve Cavalry Detachment[210]
Artillery:
1st/, 2nd/, 3rd Bn/45th Reserve Field Artillery Regt (9 btrys)
Support Troops:
45th Reserve Pioneer Company
Medical Company
45th Reserve Divisional Bridging Train

---

201 Raised in August 1914 in the 11th Corps District at Frankfurt/Main by the 81st Infantry Regiment. Prior to coming to the 44th Landwehr Division the regiment had, since the beginning of the war, been assigned to the 43rd Mixed Landwehr Brigade, the 39th Reserve Division, the Bavarian Ersatz Division, and the 54th Division.

202 According to *Stein* the 4/2nd Uhlan Regiment – designated 3rd Field Sqn as the 3rd Sqn had been converted in 1914 into the Ersatz Sqn – joined at some unspecified time from the 212th Infantry Division. According to *Voigt* that took place in 1917.

203 Raised in September 1916. Distributed in autumn 1918 among the division's infantry.

204 Joined on 22 October 1917.

The division remained in Belgium and the vicinity of Armentières through 1915 and into the first half of 1916. On 22 April it attacked in the Steenstraat sector. In 1915 the division contained:

45th Reserve Division:
  89th Reserve Brigade:
    209th Reserve Regiment
    212th Reserve Regiment
  90th Reserve Brigade:
    210th Reserve Regiment
    211th Reserve Regiment
  Attached:
    17th Reserve Jäger Battalion
Cavalry:
  45th Reserve Cavalry Detachment
Artillery:
  45th Reserve Field Artillery Regiment
Engineers
  45th Reserve Pioneer Company
  45th Reserve Pontoon Engineer
Medical:
  45th Reserve Medical Company

In the beginning of 1916 the division was north of Ypres, in Flanders, still, and remained there until 3 March. The 89th Reserve Brigade with the 209th and 212th Reserve Infantry Regiments was temporarily detached between late January and early March and assigned to the 26th Division near Becelaere. On 12 March the division was reunited in the Messines sector, south of Ypres. It remained there, inactive, until September. In early September it left Flanders and went to the Somme. The division was engaged in the Thiepval-Martinpuich sector from 9 to 24 September. On 15 September it withstood a British attack between Courcelette and Thiepval and it suffered heavy losses. After a short rest near Bapaume the division went to the Noyon area. It transferred the 209th Reserve Infantry regiment to the newly forming 207th Division. In early October the division was in the lines on the left bank of the Oise at Tracy le Val. In the interval, in order to fill out its depleted ranks, it had to borrow men from the Landsturm battalions of the 2nd Corps District. In 1916 the division contained:

45th Reserve Division:
  89th Reserve Brigade[211]:
    209th Reserve Regiment
    212th Reserve Regiment
90th Reserve Brigade:
  210th Reserve Regiment
  211th Reserve Regiment
Attached:
  17th Reserve Jäger Battalion

---

205 Raised in the 2nd Corps District at Stettin by the Ersatz Battalions of the 2nd Grenadier, 42nd and 54th Infantry Regiments. Joined 207th Infantry Division in September 1916.

206 Raised in the 9th Corps District at Altona near Hamburg by the Ersatz Battalion of the 75th Landwehr Regiment with contributions of the Ersatz Battalions of the 75th (ca. 850 men from Bremen), 76th (from Hamburg) and 31st (Altona) Infantry Regiments. More than 2/3 of the personnel had already served in peace.

207 Raised in the 2nd Corps District by the Ersatz Battalions of the 34th Fusilier (at Stettin), 49th (at Gnesen) and 140th (at Hohensalza) Infantry Regiments. About 2/3 of the personnel had already served in peace.

208 Raised in the 2nd Corps District by the Ersatz Bn of the 9th Grenadier Regiment (at Stargard) with two companies, each, of the Ersatz Bns of the 14th and 149th Infantry Regiments and personnel not yet called up from the Stargard, Belgard, Naugard and Bromberg District Commands. About 2/3 of the personnel had already served in peace, but just there were just 24 officers in autumn 1914! Disbanded on 9 October 1918 and used to reinforce the 397th Infantry Regiment.

209 Raised in autumn 1914 in the 3rd Corps District at Lübben by the Ersatz Detachment/3rd Jäger Battalion. Joined in July 1916 the 5th Jäger Regiment.

210 Raised on 28 August 1914 by the 2nd Corps District.

211 Joined on 27 November 1916 the 207th Infantry Division.

Cavalry:
- 45th Reserve Cavalry Detachment

Artillery:
- 45th Reserve Field Artillery Regiment

Engineers:
- 45th Reserve Pioneer Company
- 90th Reserve Pioneer Company[212]
- 245th Trench Mortar Company[213]
- 45th Reserve Divisional Bridging Train

In January 1917 the division was sent to the Sissonne Camp for three weeks of training. Its regiments were almost totally reformed. Between 24 September 1916 and 21 February 1917 the 210th Reserve Infantry Regiment would receive 79 NCOs and 1,522 men. On 12 February the division went into the lines west of Soissons. From 20 March to 10 April it was a reserve in the area north of Laon. Anticipating a French attack, on 10 April the division concentrated near Filain. Elements were sent to various positions to help withstand the attack of 16 April, which were withdrawn between 20 and 22 April, after having suffered heavy losses. The division was concentrated and reorganized north of Laon and went back to the lines on 3 May near the Oise-Aisne Canal. Its losses were so severe during a new French attack on 5 May that it was withdrawn from the front on 6 May.

The division was reorganized near Conflans. It then went to Côtes de Meuse, in the Verdun area, on 27 May. After three months there it was sent to Flanders in mid-September. On 22 September the division went into the lines at Zonnebeke as a counterattacking division. It moved about Flanders and on 12 October the British attacked it. Immediately after this attack, exhausted, the division was withdrawn. The division rested until it was sent into the lines northeast of St. Mihiel in the Verdun sector. It was near Bohain in December. After the withdrawal of its Altona regiment, the division consisted practically entirely of Pomeranians, however with a considerable proportion of elder men. In 1917 the division contained:

45th Reserve Division:
- 90th Reserve Brigade:
  - 210th Reserve Regiment
  - 211th Reserve Regiment
  - 212th Reserve Regiment
- Cavalry:
  - 45th Reserve Cavalry Detachment
- 45th Artillery Commend:
  - 45th Reserve Field Artillery Regiment
- 2nd Field Bn/21st Pioneer Battalion (becoming 345th Pioneer Battalion):
  - 6/21st Pioneer Battalion
  - 45th Reserve Pioneer Company
  - 245th Trench Mortar Company
  - 294th Searchlight Section
  - 23rd Reserve Searchlight Section
- Signals:
  - 445th Telephone Detachment
- Medical:
  - 527th Medical Company
  - 75th Reserve Field Hospital
  - 76th Reserve Field Hospital
  - 77th Reserve Field Hospital
  - ? Veterinary Hospital
- Transport:
  - 732nd Motor Transport Column

At the end of January 1918 the division went into the lines near Faye. It remained there and participated in the Somme offensive. It was withdrawn on 24 March, but some elements remained in the line and the rest of the division stood behind the lines in close support. During this period the division suffered heavy casualties. At the end of March it rested in the Montdidier area and in early April it went into the lines east of Montdidier.

---

212 At the end of the war serving with the 197th Infantry Division.

213 Raised in April 1916 from the 42nd Heavy Trench Mortar Detachment and the 151st Medium and 262nd Light Trench Mortar Sections. Distributed in autumn 1918 among the division's infantry.

It was relieved on the 18th and rested and refitted near Vouziers. On 1 June the division was in reserve northwest of Fère en Tardenois in the Aisne Region and on 3 June it reinforced the front near Chaudun. Shortly later it was relieved and rested.

When the Allied counteroffensive on the Marne was launched on 18 July the division was thrown into the lines near Montron. It suffered heavy losses in the actions that followed and was relieved on 27 July. The situation did not permit it to be rested and it went back into the lines two days later. On August 3 the division was relieved and rested near Maubeuge until the end of that month. When the Americans attacked in Alsace, the division was transferred to the vicinity of Muelheim and arrived there sometime before 3 September.

The division was withdrawn and sent to Flanders on 24 September, however was detrained underway near Vouziers. The 212th Reserve Regiment was sent to the Aire Valley, near Baulay, while the rest of the division went into the lines north of Fontaine en Dormois. On 8 October those elements of the division moved east and the whole division went into the lines northwest of Châtel Chéhéry. It was withdrawn on 25 September and sent to rest in Lorraine near Conflans. On 4 November it was back in the lines northeast of St. Mihiel and was there when the war ended. In early September the 212th Reserve Regiment received a draft from the 397th Regiment, of the disbanded 222nd Infantry Division. Another large replacement transport arrived at the division in mid-October, many men of which had succumbed to Communist propaganda. Insubordination and desertion to the enemy increased. In 1918 the division contained:

45th Reserve Division:
- 90th Reserve Brigade:
  - 210th Reserve Regiment
  - 211th Reserve Regiment
  - 212th Reserve Regiment
- Cavalry:
  - 45th Reserve Cavalry Detachment
  - 2/ to 5/, MG/23rd Dragoon Regiment[214]
- 45th Artillery Commend:
  - 45th Reserve Field Artillery Regiment (9 btrys)
  - 1st, 2nd & 4th Btrys, 1st Bn/20th Foot Artillery Regiment
  - 773rd Light Ammunitions Column
  - 839th Light Ammunitions Column
  - 1210th Light Ammunitions Column

345th Pioneer Battalion:
- 6/21st Pioneer Battalion
- 45th Reserve Pioneer Company
- 245th Trench Mortar Company
- ? Searchlight Section

445th Signal Command:
- 445th Telephone Detachment
- 141st Wireless Detachment

Medical:
- 527th Medical Company
- 76th Reserve Field Hospital
- 77th Reserve Field Hospital
- ? Veterinary Hospital

Transport
- 732nd Motor Transport Column

45th Landwehr Division: The 45th Landwehr Division was formed on the Eastern Front in February 1917 from Saxon troops of the 19th Corps District, when regiments and battalions of mixed Prussian-Saxon formations were sorted out. The 107th Landwehr Regiment came from the 35th Reserve Division, the 133rd Landwehr Regiment from the 92nd Division, and the 350th Landwehr Regiment from the 91st Division. A first independent 45th Landwehr Brigade had already been disbanded in December 1914. In February 1917 a second one was raised for this division. When

214 . The regiment belonged to the army cavalry, but was, according to *Stein*, was apparently attached to the division towards the end of the war.

formed the division occupied a sector in Volhynia, near the Kovel-Rovno railroad. In 1917 it contained:

45th Landwehr Division:
- 45th Landwehr Infantry Brigade:
  - 107th Landwehr Regiment
  - 133rd Landwehr Regiment[215]
  - 350th Landwehr Regiment[216]
- Cavalry:
  - 4/Saxon Guard Cavalry Regiment
- 152nd Artillery Command[217]:
  - 408th Field Artillery Regiment
- 445th Pioneer Battalion[218]:
  - 4th Reserve Company, 22nd Pioneer Battalion
  - 345th Trench Mortar Company[219]
- Signals:
  - 345th Telephone Detachment
- Medical:
  - Medical Company
  - 355th Field Hospital
  - ? Veterinary Hospital
- Transport:
  - 562nd Motor Transport Column

In February 1918 the division advanced into the Ukraine and Kiev. The Ukrainians were forming their own short-lived nation and had invited the Germans to enter their country. Though some sources indicate that the German occupation of the Ukraine was relatively unopposed, the *Histories* indicate that the 45th Landwehr Division lost heavily during its advance fighting Bolsheviks. In April the division was at Poltava and in May near Kharkov where it apparently remained until mid-October. In 1918 the division contained:

45th Landwehr Division
- 45th Landwehr Infantry Brigade:
  - 107th Landwehr Regiment
  - 133rd Landwehr Regiment
  - 350th Landwehr Regiment
- Cavalry:
  - 4/Saxon Guard Cavalry Regiment
  - 23rd Guard Dragoon Regiment[220]
- 152nd Artillery Command:
  - 498th Field Artillery Regiment
  - 1027th Light Ammunitions Column
  - 1043rd Light Ammunitions Column
- 445th Pioneer Battalion
  - 108th Pioneer Company[221]
  - 4th Reserve Company, 22nd Pioneer Battalion[222]
  - 221st Searchlight Section
- 345th Signal Command:
  - 345th Telephone Detachment
- Medical:
  - 639th Medical Company
  - 355th Field Hospital
  - 45th Veterinary Hospital

---

215 Raised in August 1914, in the 19th Corps District, for the Posen Fortress Main Reserve in Saxony by the Leipzig District Command (1st Bn), the 139th (2nd Bn; at Doebeln) and 179th (3rd Bn; at Wurzen). Joined the Bredow Landwehr Division – later 18th – and in February 1917 the 45th Landwehr Division. It may not have come from the 92nd Division. In May 1915, at Posen, the 2nd Ersatz Battalion/100th Grenadier Regiment raised a 4th Bn, which became 2nd Bn/350th Saxon Landwehr Infantry Regiment in December 1917.

216 Raised in November 1914 as 5th Landwehr Ersatz Regiment (Landwehr Brigade Ersatz Battalions Nos. 10, 11, Saxon 47 and Saxon 48) with the Menges Division with battalions of the 3rd and 19th Corps Districts. Numbered on 2 August 1915 and turned into a fully Saxon formation, on 23 December 1917, with the former 1st Bn/329th Infantry and 4th Bn/133rd Landwehr and the former 3rd Bn of the regiment itself (old 47th Landwehr Brigade Ersatz Battalion).

217 Joined on 18 July 1917.

218 According to *Stein*, over a period he does not specify, the 1st Landsturm Pioneer Company, 8th Corps District also served with this division. It later served with the Linsingen Army Group, the Kiev Army Group, the 224th Infantry Division and the 16th Landwehr Division. At the end of the war, it returned to the 45th Landwehr Division.

219 Raised in 1917. Joined in August 1918 the new 22nd Saxon Trench Mortar Battalion.

220 Seconded from the 2nd Cavalry Division for some time in 1918.

46th Reserve Division: The 46th Reserve Division was formed of troops of the 9th and 10th Corps Districts in September and October 1914 as a division of the second wave of reserve divisions and trained at the Camp Lockstedt Training Ground near Hamburg. It entered Belgium on 12 October and was engaged in the Battle of the Yser between Dixmunde and Bixschoote on 21 October. It was continually engaged until about 15 November and suffered heavy losses. The division remained in Flanders and occupied a sector near Bixschoote. In August 1914 the division contained:

46th Reserve Division:
- 91st Reserve Brigade:
  - 1st/, 2nd/, 3rd Bn/213th Reserve Regiment[223]
  - 1st/, 2nd/, 3rd Bn 214th Reserve Regiment[224]
- 92nd Reserve Brigade:
  - 1st/, 2nd/, 3rd Bn 215th Reserve Regiment[225]
  - 1st/, 2nd/, 3rd Bn 216th Reserve Regiment[226]

Attached:
- 18th Reserve Jäger Battalion[227]

Cavalry:
- 46th Reserve Cavalry Detachment[228]

Artillery:
- 1st/, 2nd/, 3rd Bn 46th Reserve Field Artillery Regiment (9 btrys)

Support Troops:
- 46th Reserve Pioneer Company
- 46th Reserve Divisional Bridging Train
- 46th Reserve Medical Company

From 22 to 27 April 1915 the division participated in the combat around Lizerne, Het-Sas, and Steenstraat. After these engagements, through the rest of 1915 and until late February 1916 the division was in the lines between Dixmude and Ypres with

---

221 This company served, according to *Stein*, with the 54th Infantry Division, while towards the end of the war the 183rd Saxon Pioneer Company joined the division.

222 According to *Stein*, at the end of the war it was with the 23rd Reserve Division. *Stein* tells that at the end of the war also the 4th Landsturm Pioneer Company, 9th Corps District belonging to the 22nd Landwehr Division was seconded to this division.

223 Raised in Schleswig-Holstein in the 9th Corps District by the Ersatz Battalions of the 85th (at Rendsburg) and 84th (at Schleswig) Infantry and 86th (at Flensburg) Fusilier Regiments – exceptionally from reservists, 100% having served in peace. Joined 207th Infantry Division in September 1916.

224 Raised in Mecklenburg in the 9th Corps District with Hanseatic and Schleswig-Holstein contributions from the Ersatz Battalions of the 90th Fusilier and 90th Reserve Regiments (1st Bn; at Rostock and Wismar), 89th Grenadier and 76th Landwehr Regiments (2nd Bn; at Schwerin) and 162nd and 163rd Infantry Regiments (3rd Bn; at Güstrow). Supposedly including volunteer students from Rostock University. When the 46th Reserve Division was disbanded on 26 July 1918 this regiment was not. A new 3rd Bn was formed from the disbanded 215th Infantry Regiment. Joined 4th Ersatz Division in August 1918.

225 Raised in the 10th Corps District by the Ersatz Battalion of the 73rd Infantry Regiment (1st and 2nd Bn; at Hanover), with contributions of the Ersatz Battalions of the 73rd Landwehr, 74th Infantry (1st Bn) and 73rd Reserve Regiments (2nd Bn), and by the Ersatz Bn/73rd Landwehr (3rd Bn; at Hildesheim) and some reservists of the Nienburg, 1st Hanover and Hildesheim District Commands. Supposedly with volunteer students from the Hanover Technical High School (= University). Disbanded on 10 August 1918 and used to form new 3rd Bs for the 362nd Infantry and 214th Reserve Regiments.

226 Raised in the 9th Corps District by the Ersatz Battalions of the 76th Infantry (half of 1st Bn; at Hamburg) and 86th Fusilier Regiments (at Flensburg, Schleswig-Holstein). Disbanded on 31 July 1918 and distributed among the 18th Infantry Division.

227 Raised in autumn 1914 in the 9th Corps District at Ratzeburg by the Ersatz Detachment/9th Jäger Battalion. Most personnel were wartime volunteers, including many students of Kiel University. Joined in July 1916 the 5th Jäger Regiment.

228 Raised on 28 August 1914 by the 9th Corps District. Disbanded on 10 August 1918.

intermittent periods of rest near Bruges. It participated in no actions of significance during this period. In 1915 the division contained:

46th Reserve Division:
- 91st Reserve Brigade:
  - 213th Reserve Regiment
  - 214th Reserve Regiment
- 92nd Reserve Brigade:
  - 215th Reserve Regiment
  - 216th Reserve Regiment
- Attached:
  - 18th Reserve Jäger Battalion

Cavalry:
- 46th Reserve Cavalry Detachment

Artillery:
- 46th Reserve Field Artillery Regiment

Engineers:
- 46th Reserve Pioneer Company
- 46th Reserve Divisional Bridging Train

Medical:
- 46th Reserve Medical Company

The division was relieved from its position north of Ypres in late February 1916 and transferred to Werwicq, and from there, on 14 March, it moved the St. Eloi sector near Messines. While there it suffered heavy casualties, beginning in early April. It rested for a short period in April and returned to the same position, remaining there from May to September when it was moved to the Somme. On 8 September it went into action between Vermandovillers and the Chaulnes railroad. It, again, suffered heavily – especially during the French attack of 17 September. The 214th Reserve Infantry Regiment was almost completely destroyed. The division went to Ham to rest, on 8 October, and returned between the 17th and 20th. In the meantime, the 213th Reserve Regiment had left for the new 207th Infantry Division. Back in the lines the division supported attacks near Ablaincourt and the Chaulnes Wood on 21 October, some of its units losing severely. It was relieved on 25 October and rested. In November and December the division occupied a sector east of Rheims in the Champagne. In 1916 the division contained:

46th Reserve Division:
- 91st Reserve Brigade[229]:
  - 213th Reserve Regiment
  - 214th Reserve Regiment
- 92nd Reserve Brigade:
  - 215th Reserve Regiment
  - 216th Reserve Regiment
- Attached:
  - 18th Reserve Jäger Battalion
- Cavalry:
  - 46th Reserve Cavalry Detachment

Artillery:
- 46th Reserve Field Artillery Regiment

Engineers:
- 46th Reserve Pioneer Company
- 1st Reserve Company, 23rd Pioneer Battalion
- 246th Trench Mortar Company[230]
- 46th Reserve Divisional Bridging Train

Attached:
- 60th Anti-Aircraft Section
- 3rd Balloon Squadron

In January 1917 the division rested in Lorraine. On 28 January it moved to Moulin sous Touvent, arriving on 30 January and staying until 18 March. In April and May it was in the St Gobain sector, leaving it on 20 May and resting near Marle. It was reorganized and received reinforcements – the 216th Reserve Infantry Regiment received 500 men from the depot of the 76th Reserve Regiment.

---

229 Joined on 26 September 1916 the 19th Landwehr Division.

230 Raised in October 1915 from the 39th Heavy, two thirds of the 195th Medium and the 268th Light Trench Mortar Detachments. Joined in September 1918 the new 96th Pioneer Battalion – a gas battalion.

On 13 June the division went into the lines at Chemin des Dames and participated in the 22 June attack on the Epine de Chevregny-Royère Farm and then on 8 July on Panthéon-Froidmont. These two battles cost it heavy casualties, which were rebuilt using men drawn from the 94th Division, then in Russia.

On 24 July the division left the Laon Front and rested near Montmedy. On 12 August it went to Spinecourt and was placed in reserve on the right bank of the Meuse during the French attack of 20 August.

Without a break, the division then went to the Verdun Region and on 22 August was engaged at the Fosse Wood-Chaume Wood Region. It lost heavily in an attack launched on 26 August and in its counterattack on Beaumont. The division was relieved on 26 August and sent to rest and reorganized near Sedan-Longuyen until September. From 3 October to 10 November the division was again in the Fosse Wood-Chaume Wood Region. On 8 November elements of the division launched an attack, suffering heavy casualties in the process. During October and November the division experienced a dysentery epidemic. From 10 November to 15 December it rested near Longwy. In mid-December the division went to Lorraine and remained there until April 1918. While the division originally had a homogeneous Northern German composition – personnel from the 9th and 10th Corps Districts of Schleswig-Holstein[231], Hamburg, Luebeck, Mecklenburg and Hanover –, in 1917 reinforcements were sent from the Eastern Front or, during the sojourn in Lorraine, from the field recruit depot of the Upper Rhinish 99th Infantry Regiment and had been rather mixed: Poles, and men from the Pomeranian 2nd, Brandenburg 3rd and Silesian 5th Corps Districts. In 1917 the division contained:

46th Reserve Division:
- 92nd Reserve Brigade:
  - 214th Reserve Regiment
  - 215th Reserve Regiment
  - 216th Reserve Regiment
- Cavalry:
  - 46th Reserve Cavalry Detachment
- 46th Artillery Command:
  - 46th Reserve Field Artillery Regiment
- 346th Pioneer Battalion:
  - 46th Reserve Pioneer Company
  - 246th Trench Mortar Company
  - 23rd Reserve Searchlight Section
- Signals:
  - 446th Telephone Detachment
- Medical:
  - 528th Medical Company
  - 335th Field Hospital
  - 78th Reserve Field Hospital
  - ? Veterinary Hospital
- Transport:
  - 733rd Motor Transport Column
- Attached:
  - 214th Reserve Regimental Motor Transport Column

The division rested in Lorraine until about 25 April, when it was transferred to the Montdidier area, where it would remain, resting and rebuilding, until 10 June. On 10 June it reinforced the Montdidier-Noyon battle line. During the subsequent attacks the division suffered severe losses. About 2 July it was withdrawn and rested behind the Noyon Front. On 20 July it went into the lines near Buzancy, south of Soissons, but within a week it was withdrawn. The division was greatly reduced in strength because it had been unable to obtain new drafts. So, it was disbanded on 26 July 1918. The 214th

231 The *Histories* claims that after the 213th Reserve Regiment recruited in Schleswig-Holstein had left, the division ceased to be a Prussian one. That is of course nonsense as the 215th continued to recruit in Hanover and the 216th in Schleswig – both Prussian provinces – and even troops raised at Hamburg, Lübeck or in Mecklenburg were contingents of those states within the Prussian Army.

Reserve Regiment was transferred to the 4th Ersatz Division while the 215th and 216th Reserve Regiments were disbanded and distributed, the latter among the 18th Infantry Division. Prior to dissolution the division contained:

46th Reserve Division
- 92nd Reserve Brigade[232]:
  - 214th Reserve Regiment
  - 215th Reserve Regiment
  - 216th Reserve Regiment
- Cavalry:
  - 46th Reserve Cavalry Detachment
- 46th Artillery Command:
  - 46th Reserve Field Artillery Regiment
- 346th Pioneer Battalion[233]:
  - 46th Reserve Pioneer Company
  - 5th Bavarian Landwehr Pioneer Company[234]
  - 23rd Reserve Searchlight Section
  - 246th Trench Mortar Company
- Signals:
  - 446th Telephone Detachment
- Medical:
  - 528th Medical Company
  - 335th Field Hospital
  - 78th Reserve Field Hospital
  - ? Veterinary Hospital
- Transport
  - 733rd Motor Transport Column

46th Landwehr Division: In March/April 1917 the Saxon 101st and 103rd Landwehr Regiments, and the staff of the 46th Landwehr Brigade, originally the Pfeil Brigade, all originating in the 12th Corps District, were reassigned from the 14th Landwehr Division to the newly forming 46th Landwehr Division on the Eastern Front. The 33rd Landsturm Battalion was also absorbed in the process. The division then occupied the Smorgoni-Lake Narotch sector until the beginning of 1918. In September 1917 the Saxon 105th Landwehr Regiment was added to the division. In December a large number of young men of the division were drafted into the 40th Infantry Division as reinforcements before it was sent to France. The division contained in 1917:

46th Landwehr Division:
- 46th Landwehr Infantry Brigade:
  - 101st Landwehr Regiment
  - 103rd Landwehr Regiment
  - 105th Landwehr Regiment[235]
- Cavalry:
  - 4/17th Uhlan Regiment
- 140th Artillery Command:
  - 246th Field Artillery Regiment
- 446th Saxon Pioneer Battalion[236]:
  - 346th Trench Mortar Company[237]
- Signals:
  - 546th Telephone Detachment
- Medical:
  - 638th Medical Company
  - 315th Field Hospital
  - 46th Veterinary Hospital
- Transport:
  - ? Motor Transport Column

---

232 Became an independent brigade staff on 11 August 1918.

233 According to *Stein*, also the 6/18th Pioneer Battalion served for some time with this division before joining the 306th Pioneer Battalion. The 346th Pioneer Battalion served as army troops with the 17th Army at the end of the war and included the 46th Reserve Pioneer Company.

234 The former 1st Landwehr Pioneer Company, 3rd Bavarian Corps District. According to *Stein*, it joined from the 1st Bavarian Landwehr Division and left for the 30th Bavarian Reserve Division. The *Histories*, though, reports this company already for 1917 to serve in the 30th Bavarian Reserve Division.

235 This Saxon regiment was raised by the Supreme Commander East on 15 February 1917 in the process of sorting out Prussian and Saxon battalions regimented during the emergencies of 1914/1915 with the former 1st Bn/381st (originally Mobile Ersatz Bn/100th Grenadier Regiment, from the 12th Corps District), 3rd Bn/345th (originally 1st Mobile Ersatz Bn/104th Reserve

In January 1918 the division was on the Volhynian front. In February it went into the Ukraine, leaving elements behind that were to gather up material that had been abandoned by the Russians as they withdrew. In mid-March the division was north of Mohilev. The 103rd Landwehr Regiment was at the Berezina in April. The 101st Landwehr Regiment appears to have been near Minsk in early May. The entire division was, apparently, drawn together in the Minsk Region in early May. It appears to have remained there until the end of the war. In 1918 the division contained:

46th Landwehr Division:
- 46th Landwehr Infantry Brigade:
  - 101st Landwehr Regiment
  - 103rd (?) Landwehr Regiment
  - 105th Landwehr Regiment
- Cavalry:
  - 4/17th Uhlan Regiment

140th Artillery Command[238]:
- Staff/246th Field Artillery Regiment

546th Signal Command:
- 346th Telephone Detachment (1st section detached)

Medical:
- 638th Medical Company
- 46th Veterinary Hospital

47th Reserve Division: The 47th Reserve Division was another one of the second wave of reserve divisions formed in September and October 1914 from troops raised in the 7th Corps District with contributions of the 4th. The newly formed division concentrated at Metz on 20 October. It went into action at Magnaville in the Woevre Region on 31 October and at Maucourt from 6 to 11 November. On 23 December the division was shifted east to the Eastern Front. It arrived in Cracow in early December and fought at Neu-Sandec/Dunajec near Tarnow in Galicia on 8 December and suffered a serious check on 20 December. In August 1914 the division contained:

47th Reserve Division[239]:
- 109th Reserve Ersatz Brigade:
  - 217th Reserve Regiment[240]
  - 218th Reserve Regiment[241]
- 110th Reserve Ersatz Brigade:
  - 219th Reserve Regiment[242]
  - 220th Reserve Regiment[243]

Attached:
- 19th Reserve Jäger Battalion[244]

Cavalry:
- 47th Reserve Cavalry Detachment[245]

Artillery:
- 47th Reserve Field Artillery Regiment (9 btrys)

Engineers:
- 47th Reserve Pioneer Company

Regiment, from the 19th Corps District) and 3rd Bn/374th (originally 1st Mobile Ersatz Bn/133rd Reserve Regiment, from the 19th Corps District) Infantry Regiments.

236 According to *Stein*, this staff later joined the 19th Army and was apparently replaced by the 303rd Pioneer Battalion joining from the 3rd Reserve Division. The 1st Landwehr Pioneer Company, 12th Corps District and the 4th Landwehr Pioneer Company, 9th Corps District were for some time *Stein* does not tell assigned to this division. The former came from the 84th Infantry Division and would serve here at the end of the war. The latter passed to the 22nd Landwehr Division, but was seconded at the end of the war to the 45th Landwehr Division.

237 Raised in April 1917. In August 1918 it joined the new 22nd Saxon Trench Mortar Battalion.

238 Left for the 96th Infantry Division on 16 April 1918.

239 Brigades renamed 93rd and 94th Reserve Brigades on 12 December 1914.

240 Raised in the 4th Corps District by the Ersatz Battalions of the 26th (1st Bn; at Magdeburg), 66th (2nd Bn; at Magdeburg) and 27th (3rd Bn; at Halberstadt) Infantry Regiments. Joined 225th Infantry Division in November 1916. Disbanded in September 1918 and distributed among the 13th Infantry Division.

241 Raised at Minden by the Ersatz Bn/15th Infantry Regiment with contributions from the Ersatz battalions of other regiments of the 7th Corps District. Disbanded on 3 August 1918 and distributed among the 14th Reserve Division.

On 10 January 1915 the division was on the Dunajec-Gorlice front. From late January to April it was in the lines west of Tarnow, near the Tarnow-Cracow railroad. It participated in the spring and summer offensives. From 20 July to 9 August it advanced from the Wysnica to the Wieprz Rivers, reaching the Bug on 19 October, the Jaselda River on 8 September and Slonim on the 13th. In late September it was near Baranovitchi and on 19 October it was in the lines near Lipsk. In 1915 the division contained:

47th Reserve Division:
- 93rd Reserve Brigade:
  - 217th Reserve Regiment
  - 218th Reserve Regiment
- 94th Reserve Brigade:
  - 219th Reserve Regiment
  - 220th Reserve Regiment
- Attached:
  - 19th Reserve Jäger Battalion
- Cavalry:
  - 47th Reserve Cavalry Detachment
- Artillery:
  - 47th Reserve Field Artillery Regiment
- Engineers:
  - 47th Reserve Pioneer Company
  - 47th Reserve Divisional Bridging Train
- Medical:
  - 47th Reserve Medical Company

The division spent all of 1916 in the Lipsk-Baranovitchi sector and did not leave until May 1917, when it went to France. The 217th Reserve Regiment was detached from the 47th Reserve Division on 23 July 1916 to form the 225th Division. Apparently, according to *Cron*, from 7 July to 25 October 1916 also the 401st Infantry Brigade was attached. In 1916 the division contained:

47th Reserve Division:
- 93rd Reserve Brigade[246]:
  - 217th Reserve Regiment
  - 218th Reserve Regiment
- 94th Reserve Brigade:
  - 219th Reserve Regiment
  - 220th Reserve Regiment
  - Württemberg 13th Landsturm Regiment[247]
- Attached:
  - 19th Reserve Jäger Battalion
- Cavalry:
  - 47th Reserve Cavalry Detachment
- Artillery:
  - 47th Reserve Field Artillery Regiment
- Engineers:
  - 47th Reserve Pioneer Company
  - 91st Reserve Pioneer Company
  - 247th Trench Mortar Company[248]
  - 47th Reserve Divisional Bridging Train
- Attached:
  - 839th Field Artillery Battery
  - 84th Anti-Aircraft Section
  - 80th Labor Battalion

---

242 Raised in the 7th Corps District by the Ersatz Battalions of the 159th (1st Bn; at Mülheim/Ruhr) and 55th (2nd Bn; at Dortmund) Infantry Regiments and the Dortmund District Command (3rd Bn). Disbanded on 3 August 1918 and its troops sent to 159th Infantry Regiment.

243 Raised at Paderborn by the Ersatz Bn/158th Infantry Regiment with contributions of the Ersatz battalions of other regiments of the 7th Corps District. Disbanded on 3 August 1918 and its troops sent as replacements in the 14th Reserve Division.

244 Raised in autumn 1914 in the 4th Corps District at Naumburg by the Ersatz Detachment/4th Jäger Battalion. Personnel consisted of 50% wartime volunteers and Ersatz reservists not having served in peace and 50% Landsturm Jäger – personnel over the age of 39. In November 1914, 200 Landsturm Jäger were replaced by Landwehr Jäger – personnel over the age of 27 – seconded from the Ersatz Detachment/7th Jäger Battalion, 7th Corps District. Joined in May 1918 the 14th Jäger Regiment.

245 Raised on 19 September 1914 by the 4th Corps District. Was transferred according to *Stein* at some unspecified time – probably in autumn 1916 – until the end of the war to the 217th Infantry Division.

246 Became independent on 1 May 1917.

In May 1917 the division was sent to France and began arriving on 7 May. It remained in the Woevre Region until early June and went into the lines north of Braye en Laonnois on 20 June. It participated in the attacks from 22 June to 8 July, suffering heavy losses. It was engaged by the French on 23 October and then retired to Chevregny. In late October it was relieved and sent to rest. Around 20 November the division occupied the Septvaux sector in the St. Gobain Forest. After the 217th Reserve Regiment had left, the division in principle recruited in the Westphalian 7th Corps District. However, in May 1917, drafts from the Eastern Front on the 400th, 420th and 421st Infantry Regiments introduced elements from the 1st, 2nd and 3rd Corps Districts. By 1917 there was a considerable number of Poles in the division.[249] In 1917 it contained:

47th Reserve Division:
- 94th Reserve Brigade:
  - 218th Reserve Regiment
  - 219th Reserve Regiment
  - 220th Reserve Regiment
- Cavalry:
  - 4/4th Jäger zu Pferd Regiment[250]
- 47th Artillery Command:
  - 47th Reserve Field Artillery Regiment
- 347th Pioneer Battalion[251]:
  - 47th Reserve Pioneer Company
  - 91st Reserve Pioneer Company
  - 247th Trench Mortar Company
  - 335th Searchlight Section
  - 42nd Reserve Searchlight Section
- Signals:
  - 447th Telephone Detachment
- Medical:
  - 529th Medical Company
  - ? Field Hospital
  - 199th Veterinary Hospital
- Transport:
  - ? Motor Transport Column

In early February 1918 the division was withdrawn from the lines and trained for the upcoming offensive. It was engaged in the Somme offensive on 21 March near Tergnier and participated in the attack of 25 March. It was engaged again on 1 April near Lassigny and held that sector until 2 May. It then moved to the St. Quentin area and from 27 May on it followed the advance of the 113th Division. It went into the line on 1 June and was withdrawn on 20 June. The division was greatly weakened, but still it went back into the lines near Longpont without being rebuilt. At this time its companies averaged 40-50 men each. The division was disbanded on 30 June 1918. The 218th Reserve Regiment was formed into a single battalion, which became the 3rd Bn/53rd Reserve Infantry Regiment. The 219th Infantry Regiment was absorbed into the 159th Regiment. Prior to dissolution the division contained:

47th Reserve Division:
- 94th Reserve Brigade:
  - 217th Reserve Regiment
  - 219th Reserve Regiment
  - 220th Reserve Regiment
- Cavalry:
  - 4/4th Jäger zu Pferd Regiment

---

247 Raised on 15 June 1915, by the 13th Corps District, from the 24th, 25th and 26th Landsturm Battalions, 13th Corps District. Replacements were sent at first from the Württemberg Landsturm Infantry Ersatz Battalion (27th Landsturm Battalion, 17th Corps District) at the Münsingen Training Ground and later from the Ersatz Bn/126th Infantry Regiment.

248 Raised in November 1916 from the 100th Heavy, the 184th Medium and the 949th Light Trench Mortar Sections. Joined in August 1918 the new Saxon 22nd Trench Mortar Battalion.

249 They could, however, have been "indigenous" in the 7th Corps District as many Poles had moved to the Ruhr industrial and mining area.

250 Joined from the 35th Infantry Division in autumn 1916.

251 Originally numbered 38th Pioneer Battalion.

47th Artillery Command:
- 47th Reserve Field Artillery Regiment

347th Pioneer Battalion[252]:
- 47th Reserve Pioneer Company
- 91st Reserve Pioneer Company
- 42nd Reserve Searchlight Section
- 247th Trench Mortar Company

Transport:
- ? Motor Transport Column

Signals:
- 447th Telephone Detachment
- 147th Wireless Detachment

Medical:
- 529th Medical Company
- 78th Reserve Field Hospital
- 81st Reserve Field Hospital
- 82nd Reserve Field Hospital
- 199th Veterinary Hospital

47th Landwehr Division: The 47th Landwehr Division was organized in June 1916 using the previously independent Saxon 47th Mixed Landwehr Brigade. This brigade had moved into France after the Battle of the Marne and on 14 September 1914 it was in the line in the vicinity of Moronvilliers. It remained in the Champagne until early 1917. In August 1914 the brigade contained:

47th Mixed Landwehr Brigade:
- 104th Landwehr Regiment[253]
- 106th Landwehr Regiment[254]

Cavalry:
- 1st & 2nd Landwehr Squadron, 19th Corps Dist[255]

Artillery:
- Landsturm Battery, 19th Corps District

In May 1915 the brigade moved to the area north of Rheims. It was reinforced by the 113th Infantry Regiment and later by the 29th Reserve Regiment. The *Histories* does not report when those "foreign" regiments left. This force became known as the Mueller Division in October 1915. At the end of September one battalion of the 104th Landwehr Regiment was sent as a reinforcement to the 133rd Reserve Infantry Regiment, which was engaged south of Ste. Marie à Py. In 1915 the brigade contained:

Mueller Division (47th Landwehr Brigade):
- Baden 113th Infantry Regiment
- Prussian 29th Reserve Regiment[256]
- 104th Landwehr Regiment
- 106th Landwehr Regiment

Cavalry:
- 1st & 2nd Landwehr Squadron, 19th Corps District

Artillery:
- Landsturm Battery, 19th Corps District

Through 1916 the 47th Landwehr Brigade continued to occupy its sector near Rheims. In July it was organized into the 47th Landwehr Division by the addition of the Saxon 391st Infantry Regiment. In August 1916 the division contained:

47th Landwehr Division:
- 47th Landwehr Infantry Brigade:
  - 391st Infantry Regiment[257]
  - 104th Landwehr Regiment
  - 106th Landwehr Regiment
- Artillery:
  - 19th Landwehr Field Artillery Regiment[258]

---

252 The staff served at the end of the war at the Jeumont Engineer Training Ground and the two pioneer companies with the 419th Pioneer Battalion of the Duke Albrecht Army Group.

253 Raised in August 1914 in the 19th Corps District at Chemnitz.

254 Raised in August 1914 in the 19th Corps District at Plauen by the 134th Infantry Regiment and at Leipzig (3rd Bn).

255 Raised in August 1914 for the 47th Mixed Landwehr Brigade respectively by the 21st and 18th Uhlan Regiments. Both were converted – probably in 1916 – into artillery ammunition columns – which happened to most Landwehr or Landsturm Sqns disbanded in the mid-war. This is reported by *Stein*, but unknown to the *Histories*.

256 Raised in August 1914 in the 8th Corps District and originally with the 16th Reserve Division.

Cavalry[259]:
Staff/, 2/20th Hussar Regiment

2nd Landwehr Squadron, 19t h Corps District (?)

The division was withdrawn from its sector near Rheims on 23 February and sent into the lines west of La Fère as the German Army retired to the Hindenburg Line. It remained in that sector until 16 May. In mid-May the division was sent to the Eastern Front and occupied the lines in the Goroditche-Tsirin sector. In July the 391st Infantry Regiment was sent to the 219th Division in exchange for the 100th Grenadier Landwehr Regiment. By 1917 the division received replacements from both the 12th and the 19th Saxon Corps Districts In August 1917 the division contained:

47th Landwehr Division:
- 47th Landwehr Infantry Brigade:
  - 100th Grenadier Landwehr Regiment[260]
  - 104th Landwehr Regiment
  - 106th Landwehr Regiment
- Cavalry:
  - Staff/, 2/20th Hussar Regiment
- 151st Artillery Command[261]:
  - 19th Landwehr Field Artillery Regiment
- 447th Pioneer Battalion[262]:
  - 6/22nd Pioneer Battalion
  - 1st Reserve Company, 22nd Pioneer Bn
  - 347th Trench Mortar Company[263]
- Signals:
  - ? Telephone Detachment
- Medical:
  - 562nd Medical Company
  - 24th Landwehr Field Hospital
  - ? Veterinary Hospital
- Transport:
  - 104th Motor Transport Column
- Attached:
  - 136th Labor Battalion

In February 1918 the division left the Tsirin Region and joined the advance into the Ukraine. On 27 April it was between Gomel and Briansk and on 5 June in the Kiev Region. It remained in the Ukraine until the armistice was implemented. In 1918 the division contained:

---

257 Raised on 29 August 1916 by the Saxon 12th Army Corps from the 1/, 2/104th Landwehr Regiment, 4/, 11/106th Landwehr Regiment, 3/, 10/108th Schützen Regiment and 8/, 10/101st Grenadier Regiment and completed to twelve companies. Replacements were sent from the 1st Ersatz Bn/81st Infantry Regiment (from the 18th Corps District).

258 Including the former Landsturm Batteries, 12th and 19th Corps Districts. That of the 12th had originally belonged to the old 45th Mixed Landwehr Brigade disbanded in December 1914.

259 The *Histories* attributes the Landwehr Sqn, 18th Uhlans, (= 2nd Landwehr Sqn, 19th Corps District) to the division in 1916 and still in 1917. The first might be true, the second is most probably wrong as at that time it already would have been converted into a column. In 1916, the *Histories* also attributes one squadron of the 21st Uhlans. According to *Stein*, that regiment was mounted army cavalry until December 1916, when it lost its horses and fought as a Schützen regiment – even if not designated as such. No part ever was attached to this division. However, the 2/20th Hussars would join in July 1916 when that regiment was distributed by squadron.

260 Raised in August 1914 in the 12th Corps District by the Saxon 100th Grenadier Regiment at Dresden. A 4th Bn was raised by three of the regiment's companies and replacements from home in October 1916. In April 1917 it became 3rd Bn/388th Landwehr Regiment.

261 Joined on 18 July 1917 and left for the 301st Infantry Division on 10 March 1918.

262 According to *Stein*, the 447th Pioneer Battalion was renamed 44th and with five companies of that battalion with the 7th Army at the end of the war.

263 Raised in October 1916 from the 71st Trench Mortar Reinforcement. Joined in August 1918 the new 22nd Saxon Trench Mortar Battalion.

47th Landwehr Division:
- 47th Landwehr Infantry Brigade:
  - 100th Grenadier Landwehr Regiment
  - 104th Landwehr Regiment
  - 106th Landwehr Regiment
- Cavalry:
  - Staff/, 2/20th Hussar Regiment
- 151st Artillery Command:
  - 19th Landwehr Field Artillery Regiment (3rd Bn detached)
- Engineers:
  - 6/22nd Pioneer Battalion
- 547th Signal Command:
  - 547th Telephone Detachment
- Medical:
  - 562nd Medical Company
  - 24th Landwehr Field Hospital
  - ? Veterinary Hospital

48th Reserve Division: The 48th Reserve Division of the second wave of reserve divisions was formed in September and October 1914. It consisted of troops of the 18th Corps District with contributions of the 7th and 11th and trained in the Oberhofen Camp. In mid-October the division moved to the area between Armentières and La Bassée. On 1 November it went into the line at Neuve Chapelle and Wyschaete in Flanders. At the end of November the division went to the Eastern Front. On 3 December it was in Poland, near Kalisch, and near Warsaw at the end of December. In October 1914 the division contained:

48th Reserve Division:
- 95th Reserve Brigade:
  - 221st Reserve Regiment[264]
  - 222nd Reserve Regiment[265]
- 96th Reserve Brigade:
  - 223rd Reserve Regiment[266]
  - 224th Reserve Regiment[267]
- Attached:
  - 20th Reserve Jäger Battalion[268]
- Cavalry:
  - 48th Reserve Cavalry Detachment[269]
- Artillery:
  - 48th Reserve Field Artillery Regiment (9 btrys)
- Engineers:
  - 48th Reserve Pioneer Company

The division was engaged near Rawka, Poland, until 28 January 1915. On 2 February it was attached to von Linsingen's German Southern Army and fought in the

---

264 A Grand Ducal Hessian regiment within the Prussian Army raised in the 18th Corps District by the Ersatz Battalions of the 115th (1st Bn; at Darmstadt), 118th (2nd Bn; at Worms) and 168th (3rd Bn; at Offenbach) Infantry Regiments. Would supposedly contain volunteer students from the Darmstadt Technical High School (= University). In November incorporated the former 2nd Bn/118th Reserve Regiment.

265 A Grand Ducal Hessian regiment within the Prussian Army raised in the 18th Corps District by the Ersatz Battalions of the 116th (1st Bn; at Giessen), Prussian 88th (2nd Bn; at Mainz) and 117th (3rd Bn; at Mainz) Infantry Regiments. In November incorporated the former 1st Bn/118th Reserve Regiment.

266 Raised in the 18th Corps District by the Prussian Ersatz Battalions of the 81st (1st Bn; at Frankfurt/Main), 80th (2nd Bn; at Wiesbaden) and 87th (3rd Bn; at Mainz) Infantry Regiments. In November incorporated the former 3rd Bn/118th Reserve Regiment.

267 Raised in the 11th Corps District in Thuringia by the Ersatz Battalions of the 71st (1st Bn; at Erfurt), 96th (2nd Bn; at Erfurt) and 94th (3rd Bn; at Weimar) Infantry Regiments. Joined 215th Infantry Division in September 1916.

268 Raised in the 7th Corps District in autumn 1914 at Bückeburg, Schaumburg-Lippe, by the Ersatz Detachment/7th Jäger Battalion from wartime volunteers and two Ersatz companies "over the establishment" not envisaged in the mobilization plan. Joined in October 1917 the 13th Jäger Regiment.

269 Raised by the 7th Corps District on 1 September 1914 and disbanded on 30 December 1916 (personnel to the 5/1st Guard Dragoons and the 9th Cavalry Division.

Carpathians, southeast of the Beskide Mountains. From February to May it was engaged at Uzsok Ridge. The division then fought in the spring and summer offensives in Galicia, crossing the Dniester in mid-June and was between Zlota/Lipa River at the end of July. The 224th Reserve Regiment was rebuilt several times. From August to October it lost 70 officers and 4,712 men, 3,100 of whom were reported as missing. Most of those men were from Alsace and Lorraine[270] and, apparently, they had deserted. When the offensive resumed in October and November it was part of the Bothmer Army and moved as far as the Stripa River. In 1915 the division contained:

48th Reserve Division:
- 95th Reserve Brigade:
  - 221st Reserve Regiment
  - 222nd Reserve Regiment
- 96th Reserve Brigade:
  - 223rd Reserve Regiment
  - 224th Reserve Regiment
- Attached:
  - 20th Reserve Jäger Battalion
- Cavalry:
  - 48th Reserve Cavalry Detachment
- Artillery:
  - 48th Reserve Field Artillery Regiment
- Engineers:
  - 48th Reserve Pioneer Company
  - 48th Reserve Divisional Bridging Train
- Medical:
  - 48th Reserve Medical Company

In winter and spring of 1916 the division remained near the Stripa west of Tarnopol. It was still there during the Brusilov offensive from June to September. In September 1916 the 224th Reserve Infantry Regiment was transferred to the 215th Division. In early October the division went into action against Rumania and fought near Hermannstadt in Hungarian Siebenbuergen/Transylvania and in November near Preocal. Thereafter it returned to Galicia and was between Brzezan and the Dniester. In 1916 the division contained:

48th Reserve Division:
- 95th Reserve Brigade[271]:
  - 221st Reserve Regiment
  - 222nd Reserve Regiment
- 96th Reserve Brigade:
  - 223rd Reserve Regiment
  - 224th Reserve Regiment
- Attached:
  - 20th Reserve Jäger Battalion
- Cavalry:
  - 48th Reserve Cavalry Detachment
- Artillery:
  - 48th Reserve Field Artillery Regiment
- Engineers:
  - 48th Reserve Pioneer Company[272]
  - 2nd Landwehr Pioneer Company, 6th Corps District
  - 248th Trench Mortar Company[273]
  - 48th Reserve Divisional Bridging Train

In May 1917 the division was relieved and sent to France. On 29 June it was behind the Verdun Front on the left bank of the Meuse. On 30 June it underwent an artillery bombardment as the French prepared for their 17 July offensive. Elements were engaged, as reinforcements, on Hill 304 and at Mort Homme. At the end of July it was sent to Stenay to rest. While there it received a large draft of men from the Class of

270 This might be true or not. The Ersatz battalions of the active infantry regiments at home in Alsace-Lorraine were relocated to the interior including Thuringia. However, the regiment was not raised by those, but by Ersatz battalions belonging to Thuringian regiments.

271 Became independent on 14 April 1917, joined on 11 March 1918 the 12th Landwehr Division and was disbanded on 2 October 1918.

272 According to *Stein*, the 274th Pioneer Company also served for some unspecified time with this division before joining the 3rd Guard Infantry Division – apparently in 1916.

273 Raised in October 1915 from the 75th Heavy Trench Mortar Section and the 103rd Medium and the 212th Light Trench Mortar Detachments.

1918. After being rebuilt it was sent back to the Hill 304-Mort Homme sector on 20 August, when the French again attacked. The division lost heavily in this attack. It was withdrawn on 24 August and went to a construction detail until 3 September. It held a sector in the lines in Lorraine from mid-October to mid-November and then went to Alsace to rest for two months near Ensisheim. When the 224th Reserve Regiment raised in Thuringia had left in autumn 1916 the division practically became a Hessian one recruited in the Grand Duchy and the Prussian Province. During the stay on the Eastern Front, apparently, many Alsace-Lorrainers served in the division. In 1917 it contained:

48th Reserve Division:
- 96th Reserve Brigade:
  - 221st Reserve Regiment
  - 222nd Reserve Regiment
  - 223rd Reserve Regiment
- Cavalry:
  - 5/1st Guard Dragoon Regiment[274]
- 48th Artillery Command:
  - 48th Reserve Field Artillery Regiment
- 348th Pioneer Battalion:
  - 48th Reserve Pioneer Company
  - 1st Reserve Company, 26th Pioneer Battalion
  - 248th Trench Mortar Company
- Signals:
  - 448th Telephone Detachment
- Medical:
  - 530th Medical Company
  - 79th Reserve Field Hospital
  - ? Veterinary Hospital
- Transport:
  - 587th Motor Transport Column

About 1 March 1918 the division went into reserve in Alsace. On 14 April it was sent into the lines in the Vieux Berquin sector where it remained until 26/27 May. Subsequently it rested in the Lille area until 28 June. It then returned to its former sector at Vieux Berquin. On 3 July, though, it was moved to Douai. On 7 July the division was in the lines southwest of Gavrolle and remained there until 6 October. It was used reinforce the Cambrai-St. Quentin Front, near Cambrai, on 7 October and remained there until the armistice, almost constantly engaged by the British. In late October the division was reinforced by drafts on the dissolved 118th Reserve Regiment of the 25th Reserve Division. In 1918 the division contained:

48th Reserve Division:
- 96th Reserve Brigade:
  - 221st Reserve Regiment
  - 222nd Reserve Regiment
  - 223rd Reserve Regiment
- Cavalry:
  - 5/1st Guard Dragoon Regiment
- 48th Artillery Command:
  - 48th Reserve Field Artillery Regiment
  - 1st & 3rd Btrys, 1st Bn/23rd Foot Artillery Regiment
  - 752nd Light Ammunitions Column
  - 954th Light Ammunitions Column
  - 1382nd Light Ammunitions Column
- 348th Pioneer Battalion:
  - 1st Reserve Company, 26th Pioneer Battalion
  - 48th Reserve Pioneer Company
  - 248th Trench Mortar Company
  - 214th Searchlight Section
- 448th Signal Command:
  - 448th Telephone Detachment
  - 69th Wireless Detachment
- Medical:
  - 530th Medical Company
  - 102nd Field Hospital
  - 79th Reserve Field Hospital
  - 448th Veterinary Hospital
- Transport
  - 735th Motor Transport Column

---

[274] The 5th Sqn was the 1st Field Sqn, as the regiment's 1st Sqn was its Ersatz Sqn. Joined on 11 November 1916 from the 3rd Guard Cavalry Brigade, which had become independent in October 1916.

48th Landwehr Division: The 48th Landwehr Division was formed in September 1917 by grouping three Landsturm regiments which previously had been employed in calm sectors on the Lorraine Front or behind the lines. Immediately the division was sent into the lines at Avricourt. The average age of the personnel was 40 years. However, there were also young men of reduced fitness who were later sent to other formations when service had improved their physical condition. In 1917 the division contained:

48th Landwehr Division
  1st Landsturm Brigade[275]:
    38th Landsturm Regiment[276]
    39th Landsturm Regiment[277]
    47th Landsturm Regiment[278]
    1/Guard Reserve Uhlan Regiment[279]

---

275 Originally the 72nd Reserve Infantry Brigade of the 1st Reserve Division. Renamed 1st Landsturm Brigade on 16 June 1917 becoming independent. Joining the division on 13 September 1917. A previous 1st Landsturm Brigade with the 4th Line-of-Communication Inspection had already been disbanded on 25 October 1914.

276 Raised on 7 July 1916 with four battalions by the 4th Army as Kündinger Landsturm Regiment with the Hersfeld Landsturm Infantry Bn (7th Landsturm Bn, 11th Corps District; became later 1st Bn/47th Landsturm Infantry Regiment) – later exchanged for the 2nd Potsdam Landsturm Infantry Bn (13th Landsturm Bn, 3rd Corps District; coming from 39th Landsturm Infantry Regiment) –, the 2nd Spandau Landsturm Infantry Bn (22nd Landsturm Bn, 3rd Corps District), Hanau Landsturm Infantry Bn (7th Landsturm Bn, 18th Corps District; became later 3rd Bn/47th Landsturm Infantry Regiment) – later exchanged for the 1st Essen Landsturm Infantry Bn (60th Landsturm Bn, 7th Corps District; coming from 39th Landsturm Infantry Regiment) –, and the Strassburg Landsturm Infantry Bn (1st Landsturm Bn, 15th Corps District; became later 2nd Bn/47th Landsturm Infantry Regiment). The regiment was called "A" Landsturm Infantry Regiment, 4th Army on 13 August 1916 and numbered on 26 August. Replacements were sent from the (1st) Ersatz Bn/24th Infantry Regiment (at Neuruppin, 3rd Corps District).

277 Raised with four battalions by the 4th Army as Schellerer Landsturm Regiment with the Württemberg Hall Landsturm Infantry Bn (11th Landsturm Bn, 13th Corps District), the Württemberg Calw Landsturm Infantry Bn (1st Landsturm Bn, 13th Corps District; later 3rd Bn), the 2nd Potsdam Landsturm Infantry Bn (13th Landsturm Bn, 3rd Corps District; became later 1st Bn/38th Landsturm Infantry Regiment) – later exchanged for the Württemberg Horb Landsturm Infantry Bn (5th Landsturm Bn, 13th Corps District; became later 2nd Bn of the regiment) – and the 1st Essen Landsturm Infantry Bn (60th Landsturm Bn, 7th Corps District; became later 3rd Bn/38th Landsturm Infantry Regiment). The regiment subsequently was called "B" Landsturm Infantry Regiment, 4th Army, then Stawitzky Landsturm Infantry Regiment and numbered on 4 September 1916. Replacements were sent from the Ersatz Bn(122nd Fusilier Regiment (at Heilbronn, 13th Corps District). It became a Württemberg formation on 31 May 1917 and was disbanded on 12 June 1918, the battalions being employed on the lines-of-communications.

278 Raised with three battalions by the 4th Army with the Hersfeld Landsturm Infantry Bn (7th Landsturm Bn, 11th Corps District; previously the 1st Bn/38th Landsturm Infantry Regiment), the Strassburg Landsturm Infantry Bn (1st Landsturm Bn, 15th Corps District; previously the 4th Bn/38th Landsturm Infantry Regiment; became independent again on 7 September 1917) – exchanged for the Tannenberg Guard Landsturm Infantry Bn (15th Landsturm Bn, Guard Corps District) – and the Hanau Landsturm Infantry Bn (7th Landsturm Bn, 18th Corps District; previously 3rd Bn/38th Landsturm Infantry Regiment). Replacements were sent from the Frankfurt (Main) Landsturm Infantry Ersatz Bn (31st Landsturm Bn, 18th Corps District).

279 Joined according to *Stein* the 1st Landsturm Brigade and with it came to the 48th Landwehr Division. Left at some unspecified time for the Duke Albrecht Army Group – apparently in late 1918.

The division held the Delme sector through 1918. It was not engaged in any significant action. Apparently, in early 1918 the division contained:

48th Landwehr Division
- 1st Landsturm Brigade:
  - 38th Landsturm Regiment
  - 40th Landsturm Regiment[280]
  - 47th Landsturm Regiment
- Cavalry:
  - unknown
- 26th Artillery Command[281]:
  - 264th Field Artillery Regiment
- 1st Field Bn/28th Pioneer Battalion[282]:
  - 1st Landsturm Pioneer Company, 3rd Corps District
  - 455th Trench Mortar Company[283]
- Signals:
  - 590th Telephone Detachment
- Medical:
  - ? Medical Company
  - 80th Reserve Field Hospital
  - ? Veterinary Hospital
- Transport:
  - 800th Motor Transport Column
- Attached:
  - 2nd Ratisbonne Landsturm Battalion (14th Landsturm Battalion, 3rd Bavarian Corps District)

Later in 1918 the division contained:

48th Landwehr Division
- 1st Landsturm Brigade:
  - 38th Landsturm Regiment
  - 40th Landsturm Regiment
  - 47th Landsturm Regiment
- Cavalry:
  - 1/Guard Reserve Uhlan Regiment
  - 3/14th Uhlan Regiment[284]
- 26th Artillery Command
  - 264th Field Artillery Regiment
- 224th Pioneer Battalion[285]:
  - 1st Landwehr Pioneer Company, Guard Army Corps
  - 210th Searchlight Section
  - 1st Landsturm Pioneer Company, 3rd Corps District
- 548th Signal Command:
  - 548th Telephone Detachment
  - 67th Wireless Detachment
- Medical:
  - 602nd Medical Company
  - 51st Field Hospital
  - 80th Reserve Field Hospital
  - 570th Veterinary Hospital
- Transport:
  - 800th Motor Transport Column

---

280 Apparently at some unspecified time the 40th Landsturm Infantry Regiment was seconded to the brigade, perhaps even before the 39th had left. Raised on 26 September 1916 by the 4th District Command with the Naumburg Landsturm Infantry Bn (11th Landsturm Bn, 4th Corps District), the 1st Kreuznach Landsturm Infantry Bn (8th Landsturm Bn, 21st Corps District), the Saxon Wurzen Landsturm Infantry Bn (9th Landsturm Bn, 19th Corps District) and the Saxon 3rd Landsturm Infantry Ersatz Bn, 12th Corps District (12th Landsturm Bn, 12th Corps District; left on 17 February 1917). The regiment circulated from the 5th Army to the 3rd, again to the 5th, to the "A" Army Detachment, the 19th Army and the 1st Landsturm Brigade of the 48th Landwehr Division. Replacements were sent from the 4th Landsturm Infantry Ersatz Bn, 4th Corps District (22nd Landsturm Bn, 4th Corps District) at Altengrabow.

281 Joined on 18 January 1918 from the 26th Infantry Division.

282 According to *Stein*, the 1st Field Bn/28th Pioneer Battalion, reverting to the title of 28th Pioneer Battalion, served with the 1st Guard Reserve Division.

283 Joined at some time not mentioned by *Stein* from the 255th Infantry Division. Distributed in autumn 1918 among the 94th and 153rd Landwehr Regiments.

284 According to *Stein* it came from the 197th Infantry Division towards the end of the war – apparently in October 1918 when the 197th was broken up.

285 According to *Stein*, towards the end of the war, the 224th Pioneer Battalion would join from the 224th Infantry Division. *Stein* also reports the 1st Landwehr Pioneer Company, Guard Army Corps to be part of the division at the end of the war.

49th Reserve Division: The 49th Reserve Division of the second wave of reserve divisions was formed in September and October 1914 and trained in the Warthe Camp. The troops had been raised in the 5th Corps District with contributions of the 6th and 4th. On 14 October it went to East Prussia. It took part in the offensive into Poland between the Vistula and the Warta Rivers at the end of October, evaded encirclement near Lodz and fought on the Burza, Rawka, and Bolimow Fronts in December. At that time the division contained:

49th Reserve Division:
- 97th Reserve Brigade:
  - 225th Reserve Regiment[286]
  - 226th Reserve Regiment[287]
- 98th Reserve Brigade:
  - 227th Reserve Regiment[288]
  - 228th Reserve Regiment[289]
- Attached:
  - 21st Reserve Jäger Battalion[290]

Cavalry:
- 49th Reserve Cavalry Detachment[291]

Artillery:
- 49th Reserve Field Artillery Regiment (9 btrys)

Engineers:
- 49th Reserve Pioneer Company

In early January 1915 the division was engaged on the Bzura and remained there until the summer. In June the 227th Reserve Regiment joined the new 107th Infantry Division. The division participated in the Hindenburg offensive of August, entered Warsaw on 7 August and pursued the Russians in the Skierniewicz sector. It stopped its advance near Baranovitchi. In 1915 the division contained:

49th Reserve Division:
- 97th Reserve Brigade:
  - 225th Reserve Regiment
  - 226th Reserve Regiment
- 98th Reserve Brigade[292]:
  - 227th Reserve Regiment
  - 228th Reserve Regiment
- Attached:
  - 21st Reserve Jäger Battalion

Cavalry:
- 49th Reserve Cavalry Regiment[293]

Artillery:
- 49th Reserve Field Artillery Regiment

Engineers:
- 49th Reserve Pioneer Company
- 49th Reserve Divisional Bridging Train

Medical:
- 49th Reserve Medical Company

---

[286] Raised in Silesia in the 5th Corps District by the Ersatz Battalions of the 58th Infantry (1st Bn; at Glogau), 6th Grenadier (2nd Bn; at Fraustadt) and 46th Infantry Regiments (3rd Bn; at Lübben, out of area in the 3rd Corps District).

[287] The 1st and 2nd Bns were raised at Liegnitz by the Ersatz Battalion of the 7th Grenadier Regiment with contributions of those of the 19th and 47th Infantry Regiments, the 3rd Bn by the Ersatz Bn/154th Infantry Regiment (at Jauer) – all in Silesia in the 5th Corps District.

[288] Raised in the Posen Province in the 5th Corps District by the Ersatz Battalion of the 50th Infantry Regiment (at Rawitsch; 3rd Bn; at Boganow). Practically all its privates were reservists with pre-war experience. Joined the 107th Division in June 1915. Disbanded on 10 September 1918 and distributed among the 52nd and 352nd Reserve Regiments.

[289] Raised in Silesia in the 6th Corps District by the Ersatz Battalions of the 10th Grenadier (1st and 2nd Bns; at Schweidnitz) and the 23rd Infantry Regiments (3rd Bn; at Neisse).

[290] Raised in autumn 1914 in Silesia in the 5th Corps District at Hirschberg by the Ersatz Detachment/5th Jäger Battalion. Joined in autumn 1917 the 13th Jäger Regiment.

[291] Raised on 24 August 1914 by the 4th Corps District. Disbanded on 30 December 1916 (part of the personnel going to the 8th Hussars).

[292] According to *Cron* it was already renumbered by 3 June 1915 as 213th Infantry Brigade of 213th Infantry Division.

The division remained in the vicinity of Baranovitschi until July 1916. At that time it was relieved and sent to Galicia where it held the lines southwest of Brody in early October. The 225th Reserve Infantry Regiment was sent to Rumania. The division with only two regiments, the 226th and 228th, faced the Russians in the Carpathians, on the Narajowka and from December 1916 through January 1917, it fought in the Carpathians near Worochta. In 1916 the division contained:

49th Reserve Division:
- 97th Reserve Brigade:
  - 225th Reserve Regiment
  - 226th Reserve Regiment
  - 228th Reserve Regiment
- Attached:
  - 21st Reserve Jäger Battalion

Cavalry:
- 49th Reserve Cavalry Regiment (?)

Artillery:
- 49th Reserve Field Artillery Regiment

Engineers:
- 49th Reserve Pioneer Company
- 49th Reserve Divisional Bridging Train
- 249th Trench Mortar Company[294]

In January 1917 the 225th Reserve Infantry Regiment returned to the division in the Uz Valley, in Rumania. In mid-January the division was transferred to France. After resting two months near Mons and later near Lille, on 20 March it moved into the lines east of Armentières. At the end of April the division was moved to a sector near Arras in the Artois where it suffered heavy losses from 1 to 21 May. It was withdrawn and reorganized throughout June. At the end of June it went into the lines in the Steenstraat-Bixschoote sector, north of Ypres. It suffered heavy losses in the artillery preparation that preceded the Allied attack on 21 July. The division was withdrawn on 28 July, though, before the attack was launched. It rested and reorganized in August between Lille and Tournai. In September it returned to its old sector in the Artois and remained there until the end of October. It then occupied a sector south of the Ypres-Menin Road from late October to 21 November and launched an attack on 26 November in the Croisilles-Bullecourt sector (Cambrai attack). The division originally was formed by troops of the 5th and 6th Districts – Silesia and Posen. The *Histories* claims, however, that by 1917 most replacements came from the 4th Corps District – Province of Saxony, Anhalt and Altenburg. At the end of December the division was being rested near Tourcoing. In 1917 the division contained:

---

293 Combined in March 1915 from the 49th Reserve Cavalry Detachment, the 5/16th Dragoon Regiment (coming from the 17th Infantry Division or rather the Combined Dragoon Regiment, 9th Cavalry Division) and the 4/3rd Horse Grenadiers (coming from the 3rd Infantry Division). *Stein* does not say when this improvised regiment was disbanded, but that might have been in the summer or autumn of 1916, as the 5/16th Dragoons joined, in October 1916, the newly raised 93rd Infantry Division. The 4/3rd Horse Grenadiers, however, was already attached in May 1915 to the 103rd Infantry Division. *Stein* also reports the 3/16th Dragoon Regiment (coming from the 17th Infantry Division or rather the Combined Dragoon Regiment, 9th Cavalry Division) to have joined the division at some unspecified time – perhaps replacing the 4/3rd Horse Grenadiers. The 3/16th Dragoons joined apparently already in January 1916 the 84th Infantry Division.

294 Raised in 1915 from the 159th Medium Trench Mortar Detachment, the 97th Heavy and half of the 811th Light Trench Mortar Sections. Joined in autumn 1918 the 225th Reserve Infantry Regiment.

49th Reserve Division:
- 97th Reserve Brigade:
  - 225th Reserve Regiment
  - 226th Reserve Regiment
  - 228th Reserve Regiment
- Cavalry:
  - 2/2nd Guard Dragoon Regiment[295]
- 49th Artillery Command:
  - 49th Reserve Field Artillery Regiment
- 349th Pioneer Battalion:
  - 49th Reserve Pioneer Company
  - 249th Trench Mortar Company
  - 2nd Landwehr Pioneer Company, 4th Corps District[296]
- Signals:
  - 449th Telephone Detachment
- Medical:
  - 531st Medical Company
  - 85th Reserve Field Hospital
  - 155th Veterinary Hospital
- Transport:
  - ? Motor Transport Column
- Attached:
  - 49th Reserve Cyclist Company

The division was withdrawn into the reserve around 6 April 1918. On 11 April it returned to the lines and launched an attack in the Messines sector. It was engaged until 25 April. Two days later the division went into the lines south of Ypres and remained there until 2 May. It stayed behind the front until 10 May and moved out on 11 June to rest near Bruges until 10 July. It then returned to its former sector northeast of Ypres until 27 August. On 28 August it moved to a position near Douai. On 1 September the division went into the line near Fremicourt and was engaged there until 18 September. It then went to the Scarpe-Somme Region, near Cambrai, until 27 September. It lost 1,100 prisoners in September, retired from the front around 1 October and rested near Eswars. On 12 October the division went into the lines at Courcelles les Lens. It remained in this quiet sector for a week and then returned to the Ypres Front. It was last seen west of Audenarde on 2 November. In 1918 the division contained:

49th Reserve Division:
- 97th Reserve Brigade:
  - 225th Reserve Regiment
  - 226th Reserve Regiment
  - 228th Reserve Regiment
- Cavalry:
  - 2/2nd Guard Dragoon Regiment
- 49th Artillery Command:
  - 49th Reserve Field Artillery Regiment
  - 1st, 2nd & 4th Btrys, 1st Bn/25th Foot Artillery Regiment
  - 788th Light Ammunitions Column
  - 972nd Light Ammunitions Column
  - 1318th Light Ammunitions Column
- 349th Pioneer Battalion:
  - 43rd Reserve Pioneer Company
  - 49th Reserve Pioneer Company[297]
  - 249th Trench Mortar Company
  - 188th Searchlight Section
- 449th Signal Command:
  - 449th Telephone Detachment
  - 110th Wireless Detachment
- Medical:
  - 531st Medical Company
  - 83rd Reserve Field Hospital
  - 85th Reserve Field Hospital
  - 155th Veterinary Hospital

50th Infantry Division: The 50th Infantry Division was part of the series of divisions numbered with pair numbers from 50 to 58. They were formed in March 1915 as triangular divisions around surplus infantry brigades and fourth infantry regiments of existing divisions. This particular division was assembled from Westphalian formations of the 7th Corps District. The 158th Infantry Regiment was transferred from the 13th

---

295 Joined on 14 November 1917 from the 3rd Guard Cavalry Brigade which had become independent in October 1916. The *Histories* assumes the 49th Reserve Cavalry Detachment still in 1917 which is wrong as that had been disbanded by January 1917.

296 Left again for the army troops and served at the end of the war in the 3rd Infantry Division.

297 Coming from the 43rd Reserve Division.

Infantry Division, the 39th Fusilier Regiment from the 14th Reserve Division, and the 53rd Infantry Regiment from the 14th Infantry Division. The 27th Infantry Brigade, also from the 14th Infantry Division, was renamed 100th. In May 1915 the division was in the Champagne. From June to October it occupied the Tahure sector. At the end of September the division was attacked by the French and suffered 130 officers and 7,849 men as casualties. It was rested and reorganized near Vouziers from the end of October until the end of November. It then appeared east of Rheims in the Prosnes-Prunay sector. In 1915 the division contained:

50th Infantry Division:
- 100th Infantry Brigade:
  - 39th Fusilier Regiment
  - 53rd Infantry Regiment
  - 158th Infantry Regiment[298]
- 50th Field Artillery Brigade:
  - 99th Field Artillery Regiment (6 btrys)
  - 100th Field Artillery Regiment (6 btrys, 3 being howitzers)
- Cavalry:
  - unknown[299]
- Engineers:
  - 99th Pioneer Company
  - 100th Pioneer Company
- Attached:
  - 1st Landwehr Pioneer Company, 7th Corps District[300]

In April 1916 the division moved from the Prosnes-Prunay sector to a position northeast of Verdun. It went into action in early May north of Vaux, participating in the attacks on the line Caillette Wood-Damloup between 1 and 3 June, which ended with the capture of Fort Vaux by the 158th Infantry Regiment on 4 June. Afterwards the division was exhausted and sent to rest in June and July. In July the division occupied a calm sector in the Woevre Region. At the end of July the division went back to the lines at Verdun, south of Fort Vaux. It attacked on 1 August and suffered the French offensives of 8 August and 24 October. The division took heavy losses and remained in that sector until November, when it was sent to the Vauquis in the Argonnes. In 1916 it contained:

50th Infantry Division:
- 100th Infantry Brigade:
  - 39th Fusilier Regiment
  - 53rd Infantry Regiment
  - 158th Infantry Regiment
- Cavalry:
  - 1/16th Uhlan Regiment[301]
- 50th Field Artillery Brigade:
  - 99th Field Artillery Regiment
  - 100th Field Artillery Regiment
- Engineers:
  - 4/23rd Pioneer Battalion
  - 99th Pioneer Company
  - 100th Pioneer Company
  - 1st Trench Mortar Company[302]
  - 50th Trench Mortar Company[303]
  - 50th Divisional Bridging Train
- Signals:
  - 50th Telephone Detachment
- Attached:
  - 50th Cyclist Company
  - 80th Anti-Aircraft Section
  - 2nd Reserve Company, 27th Pioneer Battalion

---

298 A Lorraine regiment stationed in peace out of area in the 7th Corps District.

299 The *Histories* reports the 2nd and 3rd Sqns of an unknown Jäger zu Pferd Regiment for 1915. However *Stein* states that no part of any Jäger Regiment zu Pferd ever served with this particular division.

300 Later, according to *Stein*, joined the 23rd Landwehr Division.

301 The 1/16th Uhlan Regiment originally serving with the 13th Infantry Division joined at the end of 1915 coming from the improvised Krosigk Regiment of the 7th Cavalry Division, perhaps via the 13th Infantry Division.

302 This would only be a temporary assignment as that company belonged to the 1st Infantry Division.

The division was withdrawn from the Argonnes Front on 15 February 1917 and rested until the end of March, then went into camps at Sissonne and then Thenailles. On 15 April it went into action at Juvincourt in the Aisne Region and was attacked by the French there on 15 April. It suffered heavily and was relieved between 27 and 29 April. The division subsequently underwent reorganization near Sissone. On 10 May it went back into the lines east of Allment on the Chemin des Dames. It rested in July and at the beginning of August it returned to Chemin des Dames. Losses continued. After the French attack on La Malmaison it retired on 1 November to a position near Neuville north of the Ailette and remained there until December. In 1917 the division contained:

50th Infantry Division:
- 100th Infantry Brigade:
  - 39th Fusilier Regiment
  - 53rd Infantry Regiment
  - 158th Infantry Regiment
- Cavalry:
  - 1/16th Uhlan Regiment
- 50th Artillery Command:
  - 99th Field Artillery Regiment (9 btrys)
- 50th Pioneer Battalion:
  - 99th Pioneer Company
  - 100th Pioneer Company
  - 50th Trench Mortar Company
  - 99th Searchlight Section
- Signals:
  - 50th Telephone Detachment
- Medical:
  - 50th Medical Company
  - 337th Field Hospital
  - 338th Field Hospital
  - 340th Field Hospital
  - ? Veterinary Hospital
- Transport:
  - ? Motor Transport Column

The division was withdrawn from the lines near Ailles on 9 January 1918 and moved to Chimay, where it arrived on 14 January. It rested, rebuilt and trained there for a month. It then moved to the La Capelle-Fontenelle area training for another month. In mid-March it moved to the front and on 21 March it attacked southwest of St. Quentin. It reached Moreuil on 30 March and was withdrawn shortly afterwards because of heavy losses. The division then rested and rebuilt in the Lassigny Region.

On 27 May the division attacked near Craonne and pushed across the Aisne. It continued south until stopped by French counterattacks. On 30 April it reached Goussancourt and then the Marne east of Dormans. Again it suffered heavy losses and was withdrawn on 13 June to rest in the Laon area. On 19 July the division was thrown into the lines southwest of Rheims to counter the Allied pressure on the Chateau-Thierry salient. It was withdrawn in August. About 30 September it returned to the lines northwest of Rheims, near Prouilly and Cormicy. It was slowly driven back from that position until it reached Banogne on 7 November when it was withdrawn. It rested a day and returned to the lines on 8 November near Mezières, where it remained until the armistice was signed. In 1918 the division contained:

50th Infantry Division:
- 100th Infantry Brigade:
  - 39th Fusilier Regiment
  - 53rd Infantry Regiment
  - 158th Infantry Regiment
- Cavalry:
  - 1/16th Uhlan Regiment
- 50th Artillery Command:
  - 99th Field Artillery Regiment
  - 95th Foot Artillery Battalion

---

303 Raised in November 1915 from the 87th Heavy, the 117th Medium and the 257th Light Trench Mortar Detachments. Distributed in autumn 1918 among the division's infantry.

1178th Light Ammunitions Column
1179th Light Ammunitions Column
1204th Light Ammunitions Column
50th Pioneer Battalion[304]:
99th Pioneer Company
100th Pioneer Company
50th Trench Mortar Company

50th Signal Command:
50th Telephone Detachment
13th Wireless Detachment
Medical:
50th Medical Company
337th Field Hospital
338th Field Hospital
132nd Veterinary Hospital

50th Reserve Division: The 50th Reserve Division of the second wave of reserve divisions was formed between September and October 1914 and trained at the Alten-Grabow Training Ground within the 4th Corps District. The troops had been raised in the 4th – Prussian Province of Saxony, Anhalt and Altenburg – and Silesian 6th Corps Districts with contributions of the 5th. In mid-October it joined the 8th Army and participated in the German offensive into Poland. It fought in the battles between the Vistula and the Warta in October, at Lodz in November, now part of the 9th Army, and on the Bzura-Rawka Front in December. In October 1914 the division contained:

50th Reserve Division:
99th Reserve Brigade:
229th Reserve Regiment[305]
230th Reserve Regiment[306]
100th Reserve Brigade:
231st Reserve Regiment[307]
232nd Reserve Regiment[308]

Attached:
22nd Reserve Jäger Battalion[309]
Cavalry:
50th Reserve Cavalry Detachment[310]
Artillery:
50th Reserve Field Artillery Regiment (9 btrys)
Engineers:
50th Reserve Pioneer Company

The division operated along the Bzura during the winter and spring of 1915. In the summer it participated in Hindenburg's offensive and pursued the Russians to the vicinity of Baranovitchi. In June the 232nd Reserve Regiment was transferred to the 107th Division. In September the division was transferred to the Western Front. It arrived in

---

304 According to *Stein*, for some time towards the end of the war, the 2nd Reserve Company, 27th Pioneer Battalion also served with this division, coming from the 22nd Infantry Division and leaving for the 5th Army and finally the 206th Infantry Division.

305 Raised in the 6th Corps District, at Breslau, by the Ersatz Battalions of the 11th Grenadier and 51st Infantry Regiments. 80% volunteers – supposedly many students of the Breslau University –, only 20% already having served in peace.

306 Raised in the 6th Corps District by the Ersatz Battalions of the 157th (1st Bn; at Brieg), 63rd (2nd Bn; at Oppeln) and 22nd (at Gleiwitz) Infantry Regiments. About 50% already having served in peace.

307 Raised in the 4th Corps District at Altenburg by the Ersatz Battalion of the 153rd Infantry Regiment with contribution of other Ersatz battalions.

308 Raised in the 4th Corps District by the Ersatz Battalions of the 72nd (1st Bn; at Torgau), 36th (2nd Bn, including two companies of the Ersatz Bn/36th Landwehr Regiment; at Halle) and 93rd (at Dessau, Anhalt) Infantry Regiments and drafts on other Ersatz battalions of the 4th Corps District. Probably contained volunteer students from Halle University. Incorporated, on 10 September 1918, the 2nd Bn/227th Reserve Regiment as new 3rd Bn. Joined the 107th Infantry Division in June 1915.

309 Raised in autumn 1914 in the 6th Corps District at Oels by the Ersatz Detachment/6th Jäger Battalion with 268 reservists having already served in peace and 684 men having not. Joined in May 1918 the 14th Jäger Regiment.

310 Raised on 10 September 1914 by the 5th Corps District.

Rethel on 14 October. On 16 October it moved to the Champagne. The division cooperated in the gas attack on 27 October near Rheims and remained in the lines until 19 December. It moved to Douai on 21 December where it occupied the Roclincourt sector east of Neuville-St. Vaast in the Artois. In 1915 the division contained:

50th Reserve Division:
  99th Reserve Brigade[311]:
    229th Reserve Regiment
    230th Reserve Regiment
  100th Reserve Brigade:
    231st Reserve Regiment
    232nd Reserve Regiment
  Attached:
    22nd Reserve Jäger Battalion
Cavalry:
  50th Reserve Cavalry Regiment[312]
Artillery:
  50th Reserve Field Artillery Regiment
Engineers:
  50th Reserve Pioneer Company
  50th Reserve Divisional Bridging Train
Medical:
  50th Reserve Medical Company

The division lost heavily in fighting at Neuville-St. Vaast. It was withdrawn from the lines in March 1916 and served in a calm sector south of Armentières from April to September. On 16 September the division was relieved and sent to the Somme where it again suffered heavily. From 6 October to 11 November it occupied a sector south of the La Bassée Canal. About 20 November it returned to the Somme and remained there until 22 December, when it was withdrawn and rested near Valenciennes. While the division was originally recruited from Silesia and Prussian Saxony – 5th, 6th, and 4th Corps Districts – the *Histories* claims that after 1915 most replacements came from Lower Saxony – the 10th Corps District – completely changing the division's character. This is not necessarily true. In 1916 the division contained:

50th Reserve Division:
  99th Reserve Brigade:
    229th Reserve Regiment
    230th Reserve Regiment
    231st Reserve Regiment
  Cavalry:
    50th Reserve Cavalry Regiment (?)
Artillery:
  50th Reserve Field Artillery Regiment
Engineers:
  50th Reserve Pioneer Company
  250th Trench Mortar Company[313]
  50th Reserve Divisional Bridging Train

In January 1917 elements of the division were in reserve at Achiet le Petit. At the end of January the entire division returned to the lines at Ancre where it again lost heavily on 11 February. It left the Somme at the beginning of March and went into the lines south of Bapaume on 13 March. It then withdrew to the Hindenburg Line, near Metz en Couture, southwest of Cambrai, and remained there until 18 April.

---

311 Disbanded on 3 June 1915.

312 Combined in March 1915 from the 50th Reserve Cavalry Detachment, the 1/ and, according to *Stein*, the 2/16th Dragoon Regiment (coming respectively from the 17th Infantry Division or the Combined Dragoon Regiment, 9th Cavalry Division, and the 18th Infantry Division). *Stein* does not tell when this improvised regiment was disbanded, but it may have happened in the summer of 1916. The 1/16th Dragoons joined the 5th Ersatz Division in January 1916 and the 2/16th Dragoons the 9th Army Corps before returning in July 1916 to the 18th Infantry Division. *Voigt* reports instead the 3/16th Dragoons until July 1915 and does not mention the 2/16th Dragoons at all in this division.

313 Raised in April 1916 from the 43rd Heavy, the 160th Medium and the 265th Light Trench Mortar Sections. Joined in autumn 1918 the 229th Reserve Infantry Regiment.

After resting near Douai until mid-May the division occupied between 18 May and 10 June a sector northeast of Arras, where it did not participate in any significant actions. It was transferred to the Ypres Front about 11 June in anticipation of a British attack and was held in reserve near Roulers until 24 June. On 31 July the division went into action in St. Julien and, again, suffered heavy casualties in its efforts to take the village. It was relieved on 10 August and rested near Mons. It returned to the lines on 20 September and was engaged on the 26th near Gheluvelt. After again suffering heavy losses the division left the Ypres Front on 3 October. After 21 September the 1/231st Reserve Infantry Regiment had been reduced to 15 men and the 6/231st Reserve Infantry Regiment to 28. At the end of October the division was in the vicinity of Lille and remained there until the mid-December. It then went to Cambrai and went into the lines south of Marcoing from 21 December 1917 to 31 January 1918. In 1917 the division contained:

50th Reserve Division:
- 99th Reserve Brigade:
  - 229th Reserve Regiment
  - 230th Reserve Regiment
  - 231st Reserve Regiment
- Cavalry:
  - 50th Reserve Cavalry Detachment
- 68th Artillery Command:
  - 50th Reserve Field Artillery Regiment
- 350th Pioneer Battalion:
  - 2nd Reserve Company, 19th Pioneer Battalion
  - 50th Reserve Pioneer Company
  - 250th Trench Mortar Company
  - 325th Searchlight Section
- Signals:
  - 450th Telephone Detachment
- Medical:
  - 532nd Medical Company
  - ? Field Hospital
  - ? Veterinary Hospital
- Transport:
  - ? Motor Transport Column

The division was withdrawn from the lines south of Marcoing on 31 January and went east of Cambrai where it was trained in open warfare. On 13 March it moved out and by the 17th it was at Villers Outréaux in Picardy. On 20 March it was in the lines southwest of Cambrai and participated in the attack of 21 March. It suffered heavy losses and was relieved on the 22nd, resting near Liéramont near Péronne. On 23 March it followed the advance and by the 28th it went into the lines southwest of Albert, launching an unsuccessful attack. In its next attack, against the ridge west of Denancourt on 5 April, the division was again beaten back with heavy losses. It was relieved on 9 April and rested at Maricourt.

During the night of 17/18 April it went into the lines south of Albert and was relieved in early May. At that time it was rested and rebuilt. On 24 May it moved to Montauban and then to Carnoy to move into the lines south of Albert again. It remained there until 28 June, when it was again rested near Cambrai. On 19 July the division moved to Chauny, from where it moved into the lines near Buzancy, south of Soissons. It was forced into a gradual withdrawal north of Braine about 20 August and from there to a position between Laval and Laon. On 14 September the division was in the lines east of Soissons. When the Mennejean Farm was captured the division was driven back and suffered heavy casualties. On 2 October the division returned to the lines northeast of Rheims and was subjected to heavy fighting until the end of the war. It remained there until 11 November. In 1918 the division contained:

50th Reserve Division:
- 99th Reserve Brigade:
  - 229th Reserve Regiment
  - 230th Reserve Regiment
  - 231st Reserve Regiment
- Cavalry:
  - 50th Reserve Cavalry Detachment
- 68th Artillery Command:
  - 50th Reserve Field Artillery Regiment
  - Staff, 1st, 2nd & 3rd Btrys 81st Foot Artillery Battalion
  - 705th Light Ammunitions Column
  - 902nd Light Ammunitions Column
  - 923rd Light Ammunitions Column

350th Pioneer Battalion:
- 2nd Reserve Company, 19th Pioneer Battalion
- 50th Reserve Pioneer Company
- 250th Trench Mortar Company
- 30th Searchlight Section

450th Signal Command:
- 450th Telephone Detachment
- 60th Wireless Detachment

Medical:
- 532nd Medical Company
- 21st Reserve Field Hospital
- 4th Reserve Field Hospital
- 450th Veterinary Hospital

Transport
- 737th Motor Transport Column

51st Reserve Division: The 51st Reserve Division of the second wave of reserve divisions was raised from troops of the 11th and 8th Corps Districts with contributions from the 10th – Thuringia and Rhine, Hesse and Hanover Provinces – in September and October 1914. It went into action northeast of Ypres in mid-October and fought on the Cortemarck-Moorslede line on the 22nd and Langemarck on the 24th. It finally took up a position near Poelcappelle. In October 1914 the division contained:

51st Reserve Division:
- 101st Reserve Brigade:
  - 1st/, 2nd/, 3rd Bn/233rd Reserve Regiment[314]
  - 1st/, 2nd/, 3rd Bn/234th Reserve Regiment[315]
- 102nd Reserve Brigade:
  - 1st/, 2nd/, 3rd Bn/235th Reserve Regiment[316]
  - 1st/, 2nd/, 3rd Bn/236th Reserve Regiment[317]

Attached:
- 23rd Reserve Jäger Battalion[318]

Cavalry:
- 51st Reserve Cavalry Detachment[319]

Artillery:
- 1st/, 2nd/, 3rd Bn/51st Reserve Field Artillery Regiment (9 btrys)

Engineers:
- 51st Reserve Pioneer Company

---

314 Raised in Thuringia in the 11th Corps District by the Ersatz Battalions of the 32nd (at Meiningen) and 95th (3rd Bn; at Gotha) Infantry Regiments and joined at the Ohrdruf Training Ground. Most personnel were volunteers. Joined 195th Infantry Division in September 1916. Disbanded in May 1918. The regimental staff became the staff of the 14th Jäger Regiment, the remainder joined the 466th Infantry Regiment.

315 Raised in Prussian Electoral Hesse in the 11th Corps District by the Ersatz Battalions of the 83rd (1st Bn; at Cassel), 167th (2nd Bn; at Cassel) and 82nd (3rd Bn; at Goettingen, Hanover Province) Infantry Regiments. ¾ volunteers – most of them high school students and students of Göttingen University. A 4th Bn was raised in September 1916 and in October became 1st Bn/435th Infantry Regiment.

316 Raised in the Rhine Province, 8th Corps District, by the Ersatz Battalions of the 28th (1st Bn; at Coblenz), 68th (2nd Bn; at Coblenz) and 160th (3rd Bn; at Bonn) Infantry Regiments. Supposedly including volunteer students of Bonn University. In September 1916 a 4th Bn was raised from drafts on Landwehr regiments and in October became the 2nd Bn/435th Infantry Regiment.

317 Raised in the Rhine Province, 8th Corps District, by the Ersatz Battalions of the 25th (1st Bn; at Aachen/Aix la Chapelle), 16th (2nd Bn; at Cologne) and 65th (3rd Bn; at Cologne) Infantry Regiments. Supposedly included volunteer students of the Aachen Technical High School (= Technical University), the Cologne Business and Administration High Schools and the Cologne Medical Academy. On 31 August 1916 a 4th Bn was raised from replacements from home and in September became the 1st Bn/385th Landwehr Regiment.

The division stayed northeast of Ypres throughout 1915 and remained there until September 1916. In 1915 it contained:

51st Reserve Division:
- 101st Reserve Brigade:
  - 233rd Reserve Regiment
  - 234th Reserve Regiment
- 102nd Reserve Brigade:
  - 235th Reserve Regiment
  - 236th Reserve Regiment
- Attached:
  - 23rd Reserve Jäger Battalion
- Cavalry:
  - 51st Reserve Cavalry Detachment
- Artillery:
  - 51st Reserve Field Artillery Regiment
  - Mobile Ersatz Detachment, 26th Field Artillery Regiment
- Engineers:
  - 51st Reserve Pioneer Company
  - 51st Reserve Divisional Bridging Train
- Medical:
  - 51st Reserve Medical Company

In September 1916 the division transferred the 233rd Reserve Regiment to the newly forming 195th Division on the Eastern Front. On 16 September 1916 it was sent to the Somme and on 18 September went into action between Combles and Morval. It suffered very heavy losses from a British attack on 26 September. After spending a few days on the Lille Front it went to the Champagne in early October. There it occupied the Tahure-Rouvroy sector in mid-October. In 1916 the division contained:

51st Reserve Division:
- 101st Reserve Brigade[320]:
  - 233rd Reserve Regiment
  - 234th Reserve Regiment
- 102nd Reserve Brigade:
  - 235th Reserve Regiment
  - 236th Reserve Regiment
- Attached:
  - 23rd Reserve Jäger Battalion
- Cavalry:
  - 51st Reserve Cavalry Detachment
- Artillery:
  - 51st Reserve Field Artillery Regiment
  - Mobile Ersatz Detachment, 26th Field Artillery Regiment
- Engineers:
  - 51st Reserve Pioneer Company
  - 251st Trench Mortar Company[321]
  - 51st Reserve Divisional Bridging Train
- Attached:
  - ? Cyclist Company

The division remained on the Massiges Front in the Champagne until mid-May 1917, including a short rest in February. It was involved in the attacks on Maisons de Champagne in March and suffered heavy losses in those attacks. In mid-May it went into the lines near Nauroy and Moronvilliers, where it again lost heavily. It was withdrawn on 8 June and on 22 June returned to a sector east of Tahure. On 10 August it was withdrawn for rest. Subsequently it went to the right bank of the Meuse where it occupied the sector of Samogneux-Hill 344, about 22 August. It suffered heavy casualties in the French attack on 25 August and the German counterattack on 9 September. The division was relieved on 12 September and rested behind the Champagne front. According to the Histories, the division received by 1917 replacements from predominantly from the 11th Corps District. In 1917 it contained:

---

318 Raised in autumn 1914 in the Hanover Province in the 10th Corps District at Goslar by the Ersatz Detachment/10th Jäger Battalion. Joined in July 1916 the 5th Jäger Regiment.

319 Raised on 29 August 1914 by the 11th Corps District.

320 Joined the 195th Infantry Division on 2 August 1916.

321 Raised in May 1916 from the 59th Heavy, the 171st Medium and the 285th Light Trench Mortar Sections. Distributed in autumn 1918 among the division's infantry.

51st Reserve Division:
- 102nd Reserve Brigade:
  - 234th Reserve Regiment
  - 235th Reserve Regiment
  - 236th Reserve Regiment
- Cavalry:
  - 51st Reserve Cavalry Detachment
- 51st Artillery Command:
  - 51st Reserve Field Artillery Regiment
- 351st Pioneer Battalion:
  - 7/28th Pioneer Battalion
  - 51st Reserve Pioneer Company
  - 251st Trench Mortar Company
  - 258th Searchlight Detachment
- Signals:
  - 451st Telephone Detachment
- Medical:
  - 533rd Medical Company
  - 87th Reserve Field Hospital
  - 90th Reserve Field Hospital
  - 263rd Veterinary Hospital
- Transport:
  - 738th Motor Transport Column
- Attached:
  - ? Cyclist Company

The division was trained in the Vouziers area until 20 January 1918. It then went to the Butte du Mesnil in the Champagne. It was relieved in early March and received more training in open warfare. It participated in the Kaiserschlacht in Picardy and on 30 March the division was at Montdidier. In the night of 1/2 May it was relieved and rested. On 30 May, rested and back up to strength, it went into action near Vauxbuin in the Aisne Region. It was heavily engaged and suffered severe losses. The division was forced back by the Allied counteroffensive. It was withdrawn from the lines in early August and rested near Marle. On 28 September the division returned to the lines near Somme-Py in the Champagne. It was once more badly handled and had to be withdrawn on 4 October resting a day or two near Vouziers. From 8-15 October it occupied the Bernot-Origny sector east of St. Quentin. It rested two weeks and was sent into the lines east of Rethel on 31 October, where it remained until the armistice. In 1918 the division contained:

51st Reserve Division:
- 102nd Reserve Brigade:
  - 234th Reserve Regiment
  - 235th Reserve Regiment
  - 236th Reserve Regiment
- Cavalry:
  - 51st Reserve Cavalry Detachment
- 51st Artillery Command:
  - 51st Reserve Field Artillery Regiment
  - 1st Bn/11th Reserve Foot Artillery Regiment
  - 896th Light Ammunitions Column
  - 897th Light Ammunitions Column
  - 1395th Light Ammunitions Column
- 351st Pioneer Battalion:
  - 7/28th Pioneer Battalion
  - 51st Reserve Pioneer Company
  - 33rd Searchlight Detachment
- 451st Signal Command:
  - 451st Telephone Detachment
  - 85th Wireless Detachment
- Medical:
  - 533rd Medical Company
  - 87th Reserve Field Hospital
  - 88th Reserve Field Hospital
  - 263rd Veterinary Hospital
- Transport
  - 738th Motor Transport Column

52nd Infantry Division: The 52nd Division was formed in March 1915 by drawing the Baden 159th and 170th Infantry Regiments of the 14th Corps District from the 29th Division and the Prussian 66th Regiment of the 4th Corps District from the 7th Division. The division went into the lines south of Arras and would remain there until September 1916. When formed the division contained:

52nd Infantry Division:
104th Infantry Brigade[322]:
66th Infantry Regiment
169th Infantry Regiment
170th Infantry Regiment
Cavalry:
unknown[323]
52nd Field Artillery Brigade:
104th Field Artillery Regiment (6 btrys, including 3 of howitzers)
103rd Field Artillery Regiment (6 btrys)
Engineers:
103rd Pioneer Company
104th Pioneer Company
Attached:
1st Potsdam Landsturm Battalion, 3rd Corps District
2nd Potsdam Landsturm Battalion, 3rd Corps District
2/24th Pioneers

During the Allied 1916 offensive on the Somme the division held the Hébuterne-Beaumont-Hamel sector from September through November. On 26 November it was withdrawn and rested near Bouchain. In 1916 the division contained:

52nd Infantry Division: Generalleutnant von Borries
104th Infantry Brigade:
66th Infantry Regiment
2 Machine Gun Companies
169th Infantry Regiment
2 Machine Gun Companies
170th Infantry Regiment
2 Machine Gun Companies
74th Machine Gun Sharpshooter Troop
Cavalry:
4/16th Uhlan Regiment[324]
52nd Field Artillery Brigade:
103rd Field Artillery Regiment (6 btrys)
104th Field Artillery Regiment (6 btrys, including 3 of howitzers)
108th Flak Section
Engineers:
103rd Pioneer Company
104th Pioneer Company
52nd Trench Mortar Company[325]
103rd Searchlight Platoon
52nd Divisional Bridging Train
Signals:
52nd Telephone Detachment
158th Fortress Signal Troop
159th Fortress Signal Troop
Attached:
52nd Bicycle Company
52nd Medical Company
Field Recruit Depot

On 13 January 1917 the division was sent to southern Alsace. In mid-January it occupied the Altkirch sector and held it until late March. It rested at the beginning of April and moved to the Aisne at the end of that month. On 6 April 1917 the 66th Infantry Regiment was replaced by the Baden 111th Infantry Regiment, which came from the 28th Division. Also in April the 104th Infantry Brigade Staff went to the 228th Infantry Division and was replaced by the 56th Infantry Brigade Staff originally of the 28th Infantry Division, but now coming from the 52nd Reserve Division. On 21 April the division went into the lines south of Juvincourt. It remained there until 10 July with a break between 1 and 15 June. About 24 July the division went into the lines on the Californie Plateau and launched an attack on 14 September. It rested near Sissonne at the end of September and reappeared near Pinon in mid-October. It went into action and lost

322 Formerly the 84th Infantry Brigade of the 29th Division. Renumbered 104th on 4 March 1915. Joined 228th Infantry Division on 10 April 1917.

323 The *Histories* reports the 2/16th Uhlans during 1915 and 1916. According to *Stein*, though, that squadron served with the 14th Infantry Division and from late 1915 onwards with the 5th Landwehr Division.

324 Joined at the end of 1915 from the Korsigk Cavalry Regiment probably via the 13th Infantry Division.

325 Raised in November 1915 from the 27th Heavy, the 276th Light and one half of the 130th Medium Trench Mortar Detachments. Distributed in autumn 1918 among the division's infantry regiments.

heavily on 23 October. After resting near Chimay in early November it went into the lines on the Champagne Front and remained there until 15 December. From 15 December until January 1918 it rested near Vouziers. After the 66th Infantry Regiment had left, the division practically consisted only of Baden personnel, however with some recruits from the Rhine Province and Hesse-Nassau. In 1917 it contained:

52nd Infantry Division:
- 56th Infantry Brigade:
  - 111th Infantry Regiment
  - 169th Infantry Regiment
  - 170th Infantry Regiment
- Cavalry:
  - 4/16th Uhlan Regiment
- 52nd Artillery Command:
  - 104th Field Artillery Regiment (9 btrys)
- 137th Pioneer Battalion:
  - 103rd Pioneer Company
  - 104th Pioneer Company
  - 52nd Trench Mortar Company
  - 103rd Searchlight Section
- Signals:
  - 52nd Telephone Detachment
- Medical:
  - 52nd Medical Company
  - 341st Field Hospital
  - 342nd Field Hospital
  - 343rd Field Hospital
  - 205th Veterinary Hospital
- Attached:
  - 52nd Cyclist Company
  - 577th Divisional Supply Company
  - 147th Giant Periscope Section
  - 209th Field Signal Section
  - 210th Field Signal Section
  - Pigeon Loft
  - 107th Wireless Detachment
  - 57th Wireless Compass Station
  - 52nd Trench Wireless Detachment
  - Power Buzzer Section
  - 16th Machine Gun Sharpshooter Battalion
  - Staff/203rd Foot Artillery Regiment
  - Staff 3rd Bn/27th Foot Artillery Regiment
  - 1/94th Foot Artillery Battalion
  - 1/14th Foot Artillery Regiment
  - 1/4th Foot Artillery Regiment
  - 4th Naval Gun Battery
  - 237th Close Range Battery (37mm guns)
  - 44th Artillery Survey Section
  - 87th Supply Train
  - 103rd Motor Transport Column
  - 274th Motor Transport Column
  - 725th Motor Transport Column
  - 944th Motor Transport Column
  - 2nd Ammunition Train
  - 4th Ammunition Train
  - 136th Ammunition Train (new pattern)
  - 65th Supply Train
  - 663rd Supply Train
  - 665th Supply Train
  - 8th Air Squadron (protective flight)
  - 205th Air Squadron (reconnaissance flight)
  - 1/, 2/, 3/14th Labor Battalion
  - 1/, 6/56th Road Construction Battalion

After resting near Vouziers in mid-January 1918 the division went into the lines northeast of Suippes. Units not actually on the front lines were trained in open warfare. In early March the division was relieved and went to the Vouziers area where training in a war of movement continued. When the German offensive in Picardy began at the end of March it moved via Fresnoy le Grand-Holnon Wood to Fresnoy les Roye. It relieved the 28th Division near Hangest en Santerre on the 28th. It was, itself, relieved on 14 April and rested near Sedan.

While resting the division was thoroughly reconstructed. On 26 May it went into the liens near Juvincourt, east of Craonne, in the Aisne Region and participated in the attack on the next day. It crossed the Aisne near Pontavert on the 27th, the Vesle at Jonquery on the 28th, and reached Faverolles on the 29th. It arrived on the Marne on 2 June. After taking heavy losses the division was withdrawn and again rested near Sedan. On the 20 June it was reported in reserve near Tournai.

The division then was in the lines in the Avion sector from 13 July to 5 August in the Lens area. It subsequently moved to the lines south of Bailleul and on the 17th it was withdrawn. Five days later it reinforced the lines near Miraumont west of Bapaume. It was withdrawn on 4 September after losing 1,300 prisoners and rested near Courtrai. Its battalions were reduced to three companies, which were built up to strength using men drawn from the 29th Ersatz Regiment of the disbanded 223rd Division. On 28 September the division went into the lines near Exermont In the Argonnes. After heavy fighting it was driven back to Landres et St. Georges and was withdrawn from the lines about 14 October. It had lost 600 prisoners in the preceding days. On 1 November the division received large numbers of reinforcements and on 11 November it was back in the lines east of Busancy. In 1918 the division contained:

52nd Infantry Division:
- 56th Infantry Brigade:
  - 111th Infantry Regiment
  - 169th Infantry Regiment
  - 170th Infantry Regiment
- Cavalry:
  - 4/16th Uhlan Regiment
- 52nd Artillery Command:
  - 104th Field Artillery Regiment
  - Staff, 1st, 2nd, & 3rd Btrys, 84th Foot Artillery Battalion
- 137th Pioneer Battalion:
  - 103rd Pioneer Company
  - 104th Pioneer Company
  - 52nd Trench Mortar Company
  - 32nd Searchlight Section
- 52nd Signal Command:
  - 52nd Telephone Detachment
  - 46th Wireless Detachment
- Medical:
  - 52nd Medical Company
  - 341st Field Hospital
  - 342nd Field Hospital
  - 205th Veterinary Hospital
- Transport:
  - 577th Motor Transport Column

52nd Reserve Division: The 52nd Reserve Division, of the second wave of reserve divisions, was raised in September and October 1914, mostly from recruits of the Baden 14th Corps District, with contributions of the 8th, 11th and 18th Corps Districts, and trained at the Senne Training Ground in the 7th Corps District. It left for Belgium on 12 October. Its first battle was at Ypres on 22 October. It fought in October and November in the Langemarck-Passchendaele area, where it suffered heavy losses. Between 18 to 28 October the 240th Reserve Infantry Regiment lost 28 officers and 1,360 men as casualties. In October 1914 the division contained:

52nd Reserve Division[326]:
- 103rd Reserve Brigade:
  - 1st/, 2nd/, 3rd Bn/237th Reserve Regiment[327]
  - 1st/, 2nd/, 3rd Bn/238th Reserve Regiment[328]
- 104th Reserve Brigade:
  - 1st/, 2nd/, 3rd Bn/239th Reserve Regiment[329]
  - 1st/, 2nd/, 3rd Bn/240th Reserve Regiment[330]
- Attached:
  - 24th Reserve Jäger Battalion[331]
- Cavalry:
  - 52nd Reserve Cavalry Detachment[332]
- Artillery:
  - 1st/, 2nd/, 3rd Bn/52nd Reserve Field Artillery Regiment (9 btrys)
- Engineers:
  - 52nd Reserve Pioneer Company

---

326 The 238th, 239th and 240th Reserve Regiments belonged to the Baden contingent within the Prussian Army.

327 Raised in the Southern Rhine Province in the 8th Corps District by the Ersatz Battalions of the 29th and 69th Infantry Regiments at Trier. Joined 119th Infantry Division in August 1916.

328 Raised in the Baden 14th Corps District at Karlsruhe from the 3rd and 4th Ersatz Companies, each, of the Ersatz Battalions of the 109th, 112th and 143rd Infantry Regiments and completed with volunteers – supposedly including students of the Karlsruhe Technical High School (= Technical University).

329 Raised in the Baden 14th Corps District at Mannheim from the two Ersatz companies, each, of the Ersatz Battalions of the 110th Grenadier (1st Bn; at Mannheim; 422 other ranks), 113th (2nd Bn;

The 52nd Reserve Division remained on the Ypres Front throughout 1915. In April and May some of its regiments suffered heavy losses. The 240th Reserve Regiment lost another 25 officers and 1,268 men during this period. In 1915 the division contained:

52nd Reserve Division:
- 103rd Reserve Brigade:
  - 237th Reserve Regiment
  - 238th Reserve Regiment
- 104th Reserve Brigade:
  - 239th Reserve Regiment
  - 240th Reserve Regiment
- Attached:
  - 24th Reserve Jäger Battalion

Cavalry:
- 52nd Reserve Cavalry Detachment

Artillery:
- 52nd Reserve Field Artillery Regiment

Engineers:
- 52nd Reserve Pioneer Company
- 52nd Reserve Divisional Bridging Train

Medical:
- 52nd Reserve Medical Company

The division remained in the Ypres Region until 14 September 1916, when it was transferred to the Somme. The 237th Reserve Regiment was detached in August and sent to Galicia where it was used to organize the new 119th Division. The division was engaged in the Lesboeufs sector between 16 to 29 September and lost heavily to a British attack on 26 September. On 6 October 1916 the division occupied the Butte de Souain sector in the Champagne and moved to the vicinity of Tahure on 6 November. In 1916 it contained:

52nd Reserve Division:
- 103rd Reserve Brigade[333]:
  - 237th Reserve Regiment
  - 238th Reserve Regiment
- 104th Reserve Brigade:
  - 239th Reserve Regiment
  - 240th Reserve Regiment
- Attached:
  - 24th Reserve Jäger Battalion

Cavalry:
- 52nd Reserve Cavalry Detachment

Artillery:
- 52nd Reserve Field Artillery Regiment

Engineers:
- 52nd Reserve Pioneer Company
- 1st Landwehr Pioneer Company, 9th Corps District[334]
- 252nd Trench Mortar Company[335]
- 52nd Reserve Divisional Bridging Train

---

at Freiburg; 498 other ranks) and 170th (3rd Bn; at Offenburg; 365 other ranks) Infantry Regiments and completed with 1,532 recruits of the 1914 Class drawn from the Ersatz Bn/110th Grenadier Regiment. A 4th Bn was raised in September 1916 and became 3rd Bn/435th Infantry Regiment in October 1916.

330 Raised in the Baden 14th Corps District at Rastatt from the 3rd and 4th Ersatz Companies, each, of the Ersatz Battalions of the Prussian 40th Fusilier and the Baden 111th and 114th (3rd Bn; at Constance) Infantry Regiments and completed with volunteers. A 4th Bn was raised in September 1916 and became 4th Bn/435th Infantry Regiment in October 1916.

331 Raised in autumn 1914 in the 11th Corps District at Marburg by the Ersatz Detachment/11th Jäger Battalion. Joined in August 1916 the 8th Jäger Regiment.

332 Raised on 22 August 1914 by the 18th Corps District.

333 Disbanded on 16 August 1916. *Cron* writes that, instead, the 56th Infantry Brigade joined from the 28th Infantry division on 30 November 1916 and left on 4 April 1917 for the 54th Infantry Division.

334 According to *Stein*, it joined from the 8th Army and left again for the 20th Landwehr Division.

335 Raised in October 1915 from the 40th Heavy, the 133rd and one third of the 195th Medium and the 295th Light Trench Mortar Detachments. Distributed in autumn 1918 among the division's infantry.

From January to 20 April 1917 the division occupied the Tahure Front, south of Rouvroy in the Champagne. In March it suffered substantial losses, obliging an unscheduled and rapid reinforcement of men of the Class of 1918 who had only 4 months' training. In late April the division was sent to Sapigneul in the Aisne Region and went into the lines southeast of Berry au Bac around 5 May. On 7 May it was attacked by the French south of Supigneul. On 31 May it suffered heavy losses in its attack on Hill 108.

In early July 1917 the division was withdrawn and transferred to Belgium. It went into action on the Ypres-Menin road on 31 July and, in the subsequent British offensive, suffered heavy losses and again in attacks on 10 August. When relieved the 1st Bn/238th Reserve Infantry Regiment, was reduced to two companies. On 11 August the division left Belgium for the Champagne where it occupied a sector east of the Butte de Souain around 26 August and remained there until the end of December. The *Histories* claims that by 1917 recruitment was entirely Rhenish – which might not be true. In 1917 the division contained:

52nd Reserve Division:
- 104th Reserve Brigade:
  - 238th Reserve Regiment
  - 239th Reserve Regiment
  - 240th Reserve Regiment
- Cavalry:
  - 52nd Reserve Cavalry Detachment
- 69th Artillery Command:
  - 52nd Reserve Field Artillery Regiment
- 352nd Pioneer Battalion:
  - 8/28th Pioneer Battalion
  - 52nd Reserve Pioneer Company
  - 252nd Trench Mortar Company
  - 312th (?) Trench Mortar Company[336]
- Signals:
  - 452nd Telephone Detachment
- Medical:
  - 534th Medical Company
  - 88th Reserve Field Hospital
  - ? Veterinary Hospital
- Transport:
  - ? Motor Transport Column

The 52nd Reserve Cavalry Detachment was probably still present in the 52nd Reserve Division in 1918, but it is not certain. On 10 January the division was relieved from the Tahure sector and sent to Vouziers where it was trained in open warfare. In early March it returned to the lines and remained there until 23 April.

From Vouizers the division moved, on 28 April, to Mouscron. On 4 May it entered the lines in the Voormezeele sector south of Ypres. On 8 May it suffered heavy losses near Dickebusch. On 25 May it was withdrawn and rested. On 14 June the division was back in the lines south of Ypres. On 1 July it was relieved and rested southwest of Courtrai. From 6 to 26 August it was in the lines in the Kemmel sector. On 2 September the division reinforced the Bertincourt sector near Bapaume in the Cambrai Region. It suffered severe losses, including 1,000 prisoners and was relieved on 16 September. After resting and rebuilding, on 27 September it went into the lines southwest of Roulers. Here it again suffered heavy losses and was withdrawn in early October. On 14 October it returned to the lines, suffered once more heavy losses, and was withdrawn again on 25 October. It did not return to the lines before the end of the war. In 1918 the division contained:

---

336 According to *Stein*, it belonged to the 12th Landwehr Division and later the 7th Cavalry Division.

52nd Reserve Division:
- 104th Reserve Brigade:
  - 238th Reserve Regiment
  - 239th Reserve Regiment
  - 240th Reserve Regiment
- Cavalry:
  - 52nd Reserve Cavalry Detachment
- 69th Artillery Command:
  - 52nd Reserve Field Artillery Regiment
  - 51st Foot Artillery Battalion
  - 894th Light Ammunitions Column
  - 1365th Light Ammunitions Column
  - 1393rd Light Ammunitions Column
- 352nd Pioneer Battalion:
  - 8/28th Pioneer Battalion
  - 52nd Reserve Pioneer Company
  - 252nd Trench Mortar Company
  - 184th Searchlight Section
- 452nd Signal Command:
  - 452nd Telephone Detachment
  - 23rd Wireless Detachment
- Medical:
  - 534th Medical Company
  - 89th Field Hospital
  - 90th Field Hospital
  - 264th Veterinary Hospital

53rd Reserve Division: The Saxon 53rd Reserve Division of the second wave of reserve divisions was raised in September and October 1914 from troops of both Saxon corps districts. It was sent into Belgium and fought in the Battle of Ypres against the British: On 21 October it was engaged on the Poelcapelle-Becelaere front, then on 29 October at Gheluvelt, and on the Ypres-Menin road on 11 November. It suffered very heavy losses, with the 25th Reserve Jäger Battalion reduced to 225 men on 31 October and 73 on 4 November. On 25 November the 6/241st Reserve Infantry Regiment contained only 7 men who had left Germany with the regiment in September. In 1914 the division contained:

53rd Reserve Division:
- 105th Reserve Brigade:
  - 1st/, 2nd/, 3rd Bn/241st Reserve Regiment[337]
  - 1st/, 2nd/, 3rd Bn/242nd Reserve Regiment[338]
- 106th Reserve Brigade:
  - 1st/, 2nd/, 3rd Bn/243rd Reserve Regiment[339]
  - 1st/, 2nd/, 3rd Bn/244th Reserve Regiment[340]
- Attached:
  - 25th Reserve Jäger Battalion[341]
- Cavalry:
  - 53rd Reserve Cavalry Detachment[342]
- Artillery:
  - 1st/, 2nd/, 3rd Bn/53rd Reserve Field Artillery Regiment (9 btrys)
- Engineers:
  - 53rd (Saxon) Reserve Pioneer Company

---

337 Raised in the 12th Corps District at Dresden by the Ersatz Battalions of the 100th (1st Bn) and 101st (2nd Bn) Grenadier and the 177th (3rd Bn) Infantry Regiments. Disbanded on 18 September 1918 and battalions distributed among the 472nd Infantry, 100th and 101st Grenadier Regiments.

338 Raised by the in the 12th Corps District Ersatz Battalions of the 102nd (1st Bn; at Zittau), 102nd (2nd Bn; at Bautzen) and the 178th (3rd Bn; at Kamenz) Infantry Regiments with 75% volunteers. As the regiments were raised in Upper Lusatia, there would be a large proportion of the Sorben minority. Disbanded in September 1918 and battalions distributed among the 107th Infantry and 103rd Reserve Regiments and the 24th Reserve Division.

339 Raised in the 19th Corps District by the Ersatz Battalions of the 133rd (1st Bn; at Zwickau), 105th (2nd Bn; at Zwickau) and 134th (3rd Bn; at Plauen) Infantry Regiments. Reduced to two battalions, disbanded on 25 September 1918 and distributed among the 24th Reserve and 58th Infantry Division.

340 Raised in the 19th Corps District by the Ersatz Battalions of the 104th (1st Bn; at Chemnitz), 181st (2nd Bn; at Chemnitz) and 139th (3rd Bn; with additional drafts on the Ersatz Bn/106th Infantry Regiment; at Döbeln) Infantry Regiments. About 50% were reservists having already served in peace. Joined the 96th Infantry Division in May 1917 – perhaps via the 215th Division.

341 Raised in the 12th Corps District in autumn 1914 at Dresden by the Ersatz Detachment/13th Jäger Battalion mostly of wartime volunteers. Joined in August 1916 the Saxon 7th Jäger Regiment.

Through the winter of 1914/1915 the division remained north of Ypres. It participated in the Second Battle of Ypres and again suffered heavily. In June the division occupied the Wytschaete-St. Eloi sector and in mid-July returned to a position northeast of Ypres. In early October the 105th Reserve Brigade was sent to the Champagne to reinforce the lines near Tahure while the 106th Reserve Brigade occupied the Lys sector. In November the division was withdrawn, rested north of Coutrai and remained behind the front during the winter of 1915/1916. In 1915 the division contained:

53rd Reserve Division:
- 105th Reserve Brigade:
  - 241st Reserve Regiment
  - 242nd Reserve Regiment
- 106th Reserve Brigade:
  - 243rd Reserve Regiment
  - 244th Reserve Regiment
- Attached:
  - 25th Reserve Jäger Battalion

Cavalry:
- 53rd Reserve Cavalry Detachment

Artillery:
- 53rd Reserve Field Artillery Regiment (9 btrys)

Engineers:
- 53rd Reserve (Saxon) Pioneer Company
- 53rd Reserve Divisional Bridging Train

Medical:
- 53rd Reserve Medical Company

In late March 1916 the division left Flanders and went to a new position in the lines on the banks of the La Bassée Canal. In late August it marched to the Somme and was engaged southeast of Maurepas-Le Forest on 3 September when the French attacked. It suffered heavy losses in a counterattack and by repelling a new French offensive on 13 September. Between 6 and 12 September the 241st Reserve Infantry Regiment lost 12 officers and 1,502 men as casualties. On 3 September the 244th Reserve Infantry Regiment lost 400 prisoners.

The division was relieved on 14 September and went to the Loos Front near Hulluch where it served from 21 September to 5 October. On 9 October it was in the Champagne occupying the Prosnes sector east of Rheims. On 17 November the division was withdrawn and sent to Galicia on the Eastern Front. In December it was in reserve behind the Narajowka lines. In fall 1916 the 244th Reserve Infantry Regiment was transferred to the 215th Division. In 1916 the division contained:

53rd Reserve Division:
- 105th Reserve Brigade:
  - 241st Reserve Regiment
  - 242nd Reserve Regiment
- 106th Reserve Brigade[343]:
  - 243rd Reserve Regiment
  - 244th Reserve Regiment
- Attached:
  - 25th Reserve Jäger Battalion

Cavalry:
- 53rd Reserve Cavalry Detachment

Artillery:
- 53rd Reserve Field Artillery Regiment (9 btrys)

Engineers:
- 53rd Reserve (Saxon) Pioneer Company
- 53rd Reserve Divisional Bridging Train
- 253rd (Saxon) Trench Mortar Company[344]

---

[342] Raised on 1 September 1914 by the Saxon 12th Corps District. Disbanded before the end of the war according to *Stein* – at some time in autumn 1918 – on the Maubert-Fontaine Training Ground.

[343] Disbanded on 3 May 1917.

[344] Raised in October 1915 from the 41st Heavy and the 192nd Medium Trench Mortar Detachments, and half of the 147th Light Trench Mortar Section.

In January 1917 the division entered the lines where the Dniester and Narajowka Rivers meet. It remained there, relatively quietly, until mid-June. The division was relieved and rested during June before taking part in the German 20 July offensive and advancing north of the Dniester. It was withdrawn and transferred back to the Belgium in late November. In 1917 the division contained:

53rd Reserve Division:
- 105th Reserve Brigade:
  - 241st Reserve Regiment
  - 242nd Reserve Regiment
  - 243rd Reserve Regiment
- Cavalry:
  - 53rd Reserve Cavalry Detachment
- 155th Artillery Command[345]:
  - 53rd Reserve Field Artillery Regiment (9 btrys)
- 353rd Pioneer Battalion:
  - 279th Pioneer Company[346]
  - 53rd (Saxon) Reserve Pioneer Company
  - 253rd (Saxon) Trench Mortar Company
- Signals:
  - 453rd Telephone Detachment
- Medical:
  - Medical Company
  - 94th Reserve Field Hospital
  - ? Veterinary Hospital
- Transport:
  - ? Motor Transport Column

Until 12 February 1918 the division served on the Houthulst Front and rested subsequently near Bruges. On 1/2 March it went into the lines northeast of Flesquières and participated in the initial attack of 21 March in Picardy. It was withdrawn on 22 March and rested. It then served as a general headquarters reserve and on 27 March moved west, crossing the Somme. The division went into action starting when the 241st Reserve Regiment went into the lines near Moreuil the on 2 April. On the morning of 4 April the entire division attacked near the Arriese Cour Wood. By mid-April it was withdrawn from the Moreuil sector. Around 1 May the division was resting behind the Verdun Front. It occupied a quiet sector there until 11 June near Bethincourt. In mid-June the division went for more action into the lines near Soissons, in a sector north of Le Port. It was heavily engaged in July and August, until relieved on the 10th. By that date the 243rd Reserve Regiment was reduced to three companies per battalion, because of heavy casualties. After resting on the Argonnes Front, the division began to be disassembled. Elements, though, were used to face the American attacks of 29 September and 15 October. However, the reserve infantry regiments were disbanded in mid-September. In 1918, before dissolution, the division contained:

53rd Reserve Division:
- 105th Reserve Brigade:
  - 241st Reserve Regiment
  - 242nd Reserve Regiment
  - 243rd Reserve Regiment
- Cavalry:
  - 53rd (Saxon) Reserve Cavalry Detachment
- 155th Artillery Command:
  - 53rd (Saxon) Reserve Field Artillery Regiment (9 btrys)
- 353rd Pioneer Battalion[347]:
  - 53rd (Saxon) Reserve Pioneer Company[348]
  - 279th Pioneer Company[349]
  - 253rd Searchlight Detachment

345 Joined on 19 November 1917.

346 Joined, according to *Stein*, from the 36th Reserve Division. Originally, it served with the Searchlight Ersatz Battalion and the Southern Army.

347 According to *Stein*, this staff served with the Army Group of the German Crown Prince at the end of the war.

Signals:
453rd Telephone Detachment
Transport
? Motor Transport Column

Medical:
535th (?) Medical Company
92nd (Saxon) Reserve Field Hospital
453rd Veterinary Hospital

54th Infantry Division: The division was formed in March 1915 from surplus elements of existing divisions changing to the ternary organization. The 84th Infantry Regiment and the 35th Infantry Brigade, re-numbered 108th, came from the 18th Infantry Division, at home in Schleswig-Holstein in the 9th Corps District, the 27th Reserve Regiment from the 7th Reserve Division, at home in Prussian Saxony in the 4th Corps District, and the 90th Reserve Regiment from the 18th Reserve Division, raised in Mecklenburg in the 9th Corps District. The division was concentrated, in March, near Guise and in mid-April sent to the Champagne. It remained there until July 1915 and then was sent to Russia. In July and August it fought on the Narew and on the Niemen, southwest of Grodno, in September. In early October it returned to France and took up positions on the left bank of the Oise. In 1915 the division contained:

54th Infantry Division:
108th Infantry Brigade[350]:
84th Infantry Regiment
27th Reserve Regiment
90th Reserve Regiment

Cavalry:
1/17th Hussar Regiment[351]
54th Field Artillery Brigade:
107th Field Artillery Regiment (6 btrys)
108th Field Artillery Regiment (6 btrys)

From January to May 1916 the division rested near La Fère. On 14 May it occupied Hill 304, at Verdun. Around 11 September the division went to the right bank of the Meuse and advanced north of Fleury. On 24 October the French drove it back with heavy losses north of Fort Douaumont. In early November it was relieved and sent to Lorraine. Having barely had time to replace its losses, on 5 November it was again in the lines north of Flirey en Haye. In 1916 the division contained:

54th Infantry Division:
108th Infantry Brigade:
84th Infantry Regiment
27th Reserve Regiment
90th Reserve Regiment
Cavalry:
1/17th Hussar Regiment
54th Artillery Command:
108th Field Artillery Regiment
Engineers:
107th Pioneer Company
108th Pioneer Company
1st Landwehr Pioneer Company, 16th Corps District[352]

54th Trench Mortar Company[353]
107th Searchlight Section
Signals:
54th Telephone Detachment
Medical:
54th Medical Company
345th Field Hospital
348th Field Hospital
? Veterinary Hospital
Transport:
578th Motor Transport Column
Attached:
119th Labor Battalion

---

348 According to *Stein*, this company went, at some unspecified date, to the Army Group of the German Crown Prince and became 5/44th Pioneer Battalion. For some time towards the war's end, also the 4th Reserve Company, 12th Reserve Battalion, would, according to *Stein*, belong to this division before joining the 212th Pioneer Battalion.

349 Became before the end of the war 4/44th Pioneer Company of the 7th Army.

350 Originally the 35th Infantry Brigade of the 18th Infantry Division. Joined on 6 March 1915 and was renumbered as 108th Infantry Brigade.

351 Joined from the ephemeral Krosigk Cavalry Regiment of the 7th Cavalry Division in April 1915.

Through the winter of 1916/1917 the division remained in the calm Flirey sector. In mid-April it was sent behind the Champagne Front and on 21 April it went into the lines at Berry au Bac. The French unsuccessfully attacked its positions on 4 May, but inflicted heavy losses. The division lost 650 prisoners. On 10 May it was withdrawn.

The division was then sent to the east of Rheims from 15 May to 24 July and, though in the lines, it was heavily reinforced, receiving 2,000 men. On 25 July it was moved to the Ypres Region. It was in action from 5 to 19 August and suffered a British attack on 16 August. Losses were, again, severe – one company of the 90th Reserve Regiment being reduced to one officer and four other ranks. After a short rest of less than a week the division was sent to the lines on the Cambrai Front and underwent the British tank attack on 20 November, near Cambrai, where it lost 2,789 prisoners. It was immediately relieved, sent to rest in Lorraine and reorganized. Between August to November the division received more than 3,000 replacements – mostly reconvalescents of all ages. Still in 1917 the division was mixed from personnel from the Prussian Province of Saxony, Schleswig Holstein and Mecklenburg. However, the emergency intakes were sorted out, the 90th Reserve Regiment receiving all the Mecklenburg personnel as becoming to a Mecklenburg regiment and the 84th Infantry Regiment received those from the rest of the 9th Corps District. In 1917 the division contained:

54th Infantry Division:
- 108th Infantry Brigade[354]:
  - 84th Infantry Regiment
  - 27th Reserve Regiment
  - 90th Reserve Regiment
- Cavalry:
  - 1/17th Hussar Regiment
- 54th Artillery Command:
  - 108th Field Artillery Regiment
- 138th Pioneer Battalion:
  - 107th Pioneer Company
  - 108th Pioneer Company
  - 1st Landwehr Company, 16th Pioneer Battalion
  - 54th Trench Mortar Company
  - 107th Searchlight Section
- Signals:
  - 54th Telephone Detachment
- Medical:
  - 54th Medical Company
  - 345th Field Hospital
  - 348th Field Hospital
  - ? Veterinary Hospital
- Transport:
  - 578th Motor Transport Column
- Attached:
  - 119th Labor Battalion

About 17 February 1918 the division was relieved from its sector north of Nancy, however it remained in Lorraine until March training in open warfare. It moved to a position east of Douai in late March. On 2 April the division was at Rosières near Péronne. On the night of 3/4 April it moved into the lines at Moreuil and participated in the subsequent attack. It withdrew from the lines on 6 April and on 10 April returned to a sector south of Thennes for eight days.

After leaving the Thennes trenches the division moved to Cambrai to rest. From 23 May until 20 June it went into the lines near Ville sur Ancre. It subsequently rested north of the Somme until 3 August when it was engaged near Antheuil. On 1 September

---

352 According to *Stein* the hussars later served with the Bredow Landwehr Division and later with the 5th Landwehr Division. The company, however, might have left already by 1916.

353 Raised in November 1915 from the 115th Medium Trench Mortar Detachment, the 29th Heavy and one half of the 126th Medium Trench Mortar Sections. Distributed in autumn 1918 among the division's infantry regiments.

354 *Cron* writes that on 4 April 1917 also the 56th Infantry Brigade joined from the 52nd Reserve Division.

the division left its sector near Lassigny and moved to the lines northeast of Noyon. On 10 September it was relieved and rested for 15 days before returning on 25 September to the lines northeast of Hargicourt. The division then fought in the Le Cateau area until 12 October when it went into the reserves. On 2 November it was engaged at Ors and remained in combat until the signing of the armistice. In 1918 the division contained:

54th Infantry Division:
- 108th Infantry Brigade:
  - 84th Infantry Regiment
  - 27th Reserve Regiment
  - 90th Reserve Regiment
- Cavalry:
  - 1/17th Hussar Regiment
- 55th Artillery Command[355]:
  - 108th Field Artillery Regiment
  - 54th Foot Artillery Battalion
  - 746th Light Ammunitions Column
  - 955th Light Ammunitions Column
  - 966th Light Ammunitions Column

138th Pioneer Battalion:
- 107th Pioneer Company
- 108th Pioneer Company
- 97th Searchlight Section

54th Signal Command:
- 54th Telephone Detachment
- 87th Wireless Detachment

Medical:
- 54th Medical Company
- 345th Field Hospital
- 348th Field Hospital
- 54th Veterinary Hospital

Transport:
- 578th Motor Transport Column

54th Reserve Division: The 54th Reserve Division of the second wave of reserve divisions was raised in September and October 1914. It was formed of men from the Württemberg 13th Corps District with the addition of Saxon troops and trained on the 13th Corps District's Münsingen Training Ground. The division went into action on 21 October in Belgium in the Battle of Ypres. On 29 October it unsuccessfully attacked south of Gheluvelt and participated in the general attack of 11 November near Zonnebeke, where it suffered heavy casualties. From 21 to 29 November the 248th Reserve Regiment lost 32 officers and 1,395 men as casualties. In October 1914 the division contained:

54th Reserve Division:
- 107th Reserve Brigade:
  - 1st/, 2nd/, 3rd Bn/245th (Saxon) Reserve Regiment[356]
  - 1st/, 2nd/, 3rd Bn/246th (Württemberg) Reserve Regiment[357]
- 108th Reserve Brigade:
  - 1st/, 2nd/, 3rd Bn/247th (Württemberg) Reserve Regiment[358]
  - 1st/, 2nd/, 3rd Bn/248th (Württemberg) Reserve Regiment[359]

Attached:
- 26th (Saxon) Reserve Jäger Battalion[360]

Cavalry:
- 54th (Württemberg) Reserve Cavalry Det[361]

Artillery:
- 1st/, 2nd/, 3rd Bn/54th Reserve Field Artillery\ Regiment (9 btrys: 1st to 3rd: Saxon, 4th to 9th: Württemberg)

Engineers:
- 54th (Saxon) Reserve Pioneer Company

355 Apparently in January 1918, the 54th left for the 84th Infantry Division and was replaced by the 55th Artillery Command.

356 Raised in the Saxon 19th Corps District by the Ersatz Battalions of the 106th (1st Bn; at Leipzig), 107th (2nd Bn; also at Leipzig) and 179th (3rd Bn; at Wurzen) Infantry Regiments. Joined 192nd Infantry Division in June 1916.

357 Raised in the 13th Corps District by the Ersatz Battalions of the 119th (1st Bn; at Stuttgart), 120th (2nd Bn; also at Stuttgart) and 180th (3rd Bn; at Tübingen) Infantry Regiments, supposedly including volunteer students of the Tübingen University and 50% reservists having served in peace. Disbanded on 9 September 1918 and battalions sent to the 413th and 414th Infantry and 120th Reserve Regiments.

The division remained in Flanders throughout 1915. In April and May it lost heavily in the Second Battle of Ypres. During this period it occupied the Becelaere-Polygon Wood sector, between the Ypres-Menin Wood and the Roulers railroad. In 1915 the division contained:

54th Reserve Division:
- 107th Reserve Brigade:
  - 245th (Saxon) Reserve Regiment
  - 246th Reserve Regiment
- 108th Reserve Brigade:
  - 247th Reserve Regiment
  - 248th Reserve Regiment
- Attached:
  - 26th Reserve Jäger Battalion

Cavalry:
- 54th Reserve Cavalry Detachment

Artillery:
- 54th Reserve Field Artillery Regiment
- Ersatz Battalion, 59th Field Artillery Regiment

Engineers:
- 54th Reserve (Saxon) Pioneer Company
- 54th Reserve Divisional Bridging Train

Medical:
- 54th Reserve Medical Company

In late January and early February 1916 the division was withdrawn from the Ypres salient. During February and March it trained in the Beverloo camp. At the end of March the division went to the Artois where it occupied a sector south of Neuve Chapelle, but it was not engaged in any serious action while there. In June 1916 the 245th (Saxon) Reserve Regiment was detached to the newly forming 192nd Division. In late August the division was sent to the Somme and suffered heavy losses between 8 to 18 September. After a short stay in the Loos salient, in early October, it was sent to Lorraine and occupied a position in the lines near Blamont. In late November the division left Lorraine and went first behind the lines near Cambrai and then to the Verdun Region. It occupied a position on 22 December, west of Bezonvaux, after the French attack of 16 December. During 1916 the "foreign" troops left and the division became a Württemberg formation. At the beginning of 1916, though, it contained:

54th Reserve Division:
- 107th Reserve Brigade:
  - 245th (Saxon) Reserve Regiment
  - 246th Reserve Regiment
- 108th Reserve Brigade[362]:

Cavalry:
- 54th Reserve Cavalry Detachment

Artillery:
- 54th Reserve Field Artillery Regiment
- Ersatz Battalion, 59th Field Artillery Regiment

---

358 Raised in the 13th Corps District at Ulm by the Ersatz Battalions of the 123rd Grenadier (2nd Bn) and the 120th Infantry (1st Bn) and 127th (3rd Bn) Infantry Regiments with additional drafts on further Ersatz battalions of the 13th Corps District. Disbanded on 2 September 1918 and distributed among the 123rd Grenadier, 124th and 120th Infantry Regiments.

359 Raised in the 19th Corps District by the Ersatz Battalions of the 121st (1st Bn; at Ludwigsburg), 122nd (2nd Bn; at Heilbronn) and 126th (3rd Bn; at Bietigheim – peacetime garrison of the regiment had been Strassburg/Strasbourg in the Alsace) Infantry Regiments including 50% reservists having served in peace and 50% volunteers. Disbanded on 6 September 1918 and battalions sent to the 479th Infantry and 122nd Fusilier and 478th Infantry Regiments.

360 Raised in autumn 1914 in the Saxon 12th Corps District at Freiberg by the Ersatz Detachment/12th Jäger Battalion with the 1st and 2nd Companies of that detachment and the 3rd and 4th Ersatz Companies of the 182nd Infantry Regiment. There might have been student wartime volunteers joining from the Freiberg Mining Academy. Joined on 4 August 1916 the Saxon 7th Jäger Regiment.

361 Raised on 5 September 1914 by the 13th Corps District. According to *Stein*, it was transferred to the Ciney Training Ground (probably at some time in 1918) and disbanded on 2 September 1918.

362 Disbanded on 10 January 1917.

247th Reserve Regiment
248th Reserve Regiment
Attached:
26th (Saxon) Reserve Jäger Battalion

Engineers:
54th (Saxon) Reserve Pioneer Company[363]
254th (Prussian) Trench Mortar Company[364]
54th Reserve Divisional Bridging Train

The division was withdrawn from the Verdun Region around 23 January 1917 and moved to the Champagne where it occupied a sector between the Navarin Farm and the Tahure Hill. In March it seconded a battalion of the 246th Reserve Infantry Regiment to participate in the attacks on Maisons de Champagne on 27 March. In early May the division served as a reinforcement south of Moronvilliers. It was in action at Téton and Mont Haut from 15 May to 10 June, where it suffered heavy losses. The division was not rested, but sent immediately, in mid-June, to the sector west of Tahure where it remained until late August. With very little notice, the division was transferred to Hill 304 in the Verdun area and remained there from 24 August to 18 October. After its tour of duty in Verdun it was sent to Flanders where it occupied, from 29 October until March 1918, part of the Dixmude sector. After the Saxons had left in 1916, the division had become an entirely Württemberg formation. In 1917 the division contained:

54th Reserve Division:
107th Reserve Brigade:
245th Reserve Regiment
246th Reserve Regiment
247th Reserve Regiment
Cavalry:
54th Reserve Cavalry Detachment
70th Artillery Command:
54th Reserve Field Artillery Regiment
(9 btrys, Württemberg)
354th (Württemberg) Pioneer Battalion:
254th (Württemberg) Trench Mortar Company[365]
1st & 2nd Reserve Companies,
13th Pioneer Battalion[366]
312th Searchlight Section
Signals:
454th Telephone Detachment
Medical:
536th Medical Company
92nd Reserve Field Hospital
? Veterinary Hospital
Transport:
? Motor Transport Column

On 14 March 1918 the division left Dixmude and arrived in Cambrai on the 15th. It subsequently participated in Picardy in the advance of the Kaiserschlacht. It was engaged near Rosières on 25 March and Orvillers on the 26th. On 4 April, after having suffered heavy losses, the division was relieved and sent to the Somme, where it was held as a reserve, near the front until 10/11 April. It then held the sector at Ville sur Ancre until 27 May and again from 28 June until 25 August. It was withdrawn to the Mametz area after the British counterattack.

---

363 Left, according to *Stein*, for the 219th Infantry Division and served at the end of the war in the 40th Infantry Division.

364 Raised in November 1915 from the 262nd Light Trench Mortar Detachment, the 102nd Medium Trench Mortar Section and the 132nd Trench Mortar Supplement Detachment. Became the 160th Trench Mortar Company and joined the 3rd Naval Division in spring 1917.

365 Raised in April 1917 by the 27th Reserve Corps. Distributed in autumn 1918 among the 122nd Fusilier, the 123rd Grenadier, the 413th and 414th Infantry and the 120th Reserve Infantry Regiments.

366 The 1st & 2nd Reserve Companies, 13th Pioneer Battalion originally belonged, according to *Stein*, to the 28th Reserve Division, then to the 54th Reserve Division and finally towards the end of the war to 243rd Infantry Division.

In early September 1918 the division was dissolved and its regiments disbanded. In 1918, before it was dissolved, the division contained:

54th Reserve Division:
- 107th Reserve Brigade[367]:
  - 245th Reserve Regiment
  - 246th Reserve Regiment
  - 247th Reserve Regiment
- Cavalry:
  - 54th (Württemberg) Reserve Cavalry Detachment
- 70th Artillery Command:
  - 54th (Württemberg) Reserve Field Artillery Regiment
- 354th (Württemberg) Pioneer Battalion[368]:
  - 1st Reserve Company, 13th Pioneer Battalion
  - 2nd Reserve Company, 13th Pioneer Battalion
  - 312th Searchlight Section
  - 254th (Württemberg) Trench Mortar Company
- Signals:
  - 454th Telephone Detachment
- Medical:
  - 536th (Württemberg) Medical Company
  - 93rd (Württemberg) Reserve Field Hospital
  - 94th (Württemberg) Reserve Field Hospital
  - 246th Veterinary Hospital
- Transport:
  - ? Motor Transport Column

55th Division: This division was not raised.

367 Disbanded on 30 August 1918.
368 Joined the 3rd Army.

Bibliography

Cron, H.: *Geschichte des Deutschen Heeres im Weltkriege, 1914 – 1918*, Berlin, Verlag Karl Siegesmund, 1937.

Deisenroth, K.: *Immer feste druff!? – Militaer-politische Aspekte im Reichsland Elsass-Lothringen am Beispiel des XV. Armee-Korps 1871 bis 1918*. In: *Zeitschrift fuer Heereskunde* (2003)

Deutscher Generalstab: *Die Schlachten und Gefechte des Grossen Krieges 1914 – 1918*, Berlin, Verlag Hermann Sack, 1919.

Dominik, H.: *Das Matrosen-Regiment Nr. 5 1914 – 1918*. Oldenburg, 1929

Edmonds, Major, J.E.: *Handbook of the German Army, (Home and Colonial)*, London, Her Majesty's Stationary Office, 1900.

Fick, R. ed.: *Das 2. Matrosen-Regiment 1914 – 1918 in Flandern*. Luebeck, 1926

Gaebler, P.: *Das Marinekorps im Weltkrieg* 1914 – 1918. Berlin, 1926

Goetze, W. v.: *Das Marine-Infanterie-Regiment 2 im Weltkrieg 1914/18*. Oldenburg, 1926

Histories of Two Hundred and Fifty-One Divisions of the German Army Which Participated in the War (1914-1918), Washington, DC, Government Printing Office, 1920.

Huelsen, B. v.: *Das Marinekorps in Flandern 1914 – 1918*. In: *Marinerundschau* (1924)

Kling, H., Stein, H. R. v.: *Die deutschen Infanterie-Regimenter im Weltkrieg 1914 – 1918*. In: *Zeitschrift fuer Heereskunde* (1964)

Kling, H., Stein, H. R. v.: *Die Formationen der Feldartillerie 1914/1918 – Schluss*. In: *Feldgrau – Mitteilungen einer Arbeitsgemeinschaft*. Lehrte (1958)

Kling, H., Stein, H. R. v.: *Die Regimenter 383 – 479*. In: *Zeitschrift fuer Heereskunde* (1966, 1967)

Kling, H., Stein, H. R. v.: *Zur Formationsgeschichte der Infanterie 1914/18. Die im Winter 1914 zusammengestellten Regimenter 329 – 354 u. 372 – 381*. In: *Zeitschrift fuer Heereskunde* (1959)

Kling, H.: *Die Infanterie-Regimenter 183 – 190, 192, 193, w. Res. Inf. Rgt. 122 und die einzeln aufgestellten Regimenter*. In: *Zeitschrift fuer Heereskunde* (1965)

Kling, H.: *Die Jäger und Schützen (einschliesslich Bayern) und das wuerttembergische Gebirgs-Regiment.* In: *Zeitschrift fuer Heereskunde* (1962)

Kling, H.: *Die jungen Regimenter (Res. Inf. Rgt. 201 – 248., 43. – 54. Res. D., XXII. – XXVII. R. K.).* In: *Zeitschrift fuer Heereskunde* (1963)

Kling, H.: *Die Marine-Infanterie und die Matrosen-Regimenter im Weltkrieg 1914 – 1918.* In: *Feldgrau – Mitteilungen einer Arbeitsgemeinschaft.* Lehrte (1960)

Kling, H.: *Die Neuformationen der Feldartillerie 1914 – 1918.* In: *Feldgrau – Mitteilungen einer Arbeitsgemeinschaft.* Lehrte (1958)

Kling, H.: *Die Neuformationen der Kavallerie 1914/18.* In: *Feldgrau – Mitteilungen einer Arbeitsgemeinschaft.* Lehrte (1955)

Kling, H.: *Die Reserve-Regimenter 249 – 272 (75. – 82. Reserve-Division).* In: *Zeitschrift fuer Heereskunde* (1965)

Kling, H.: *Die Truppenteile der deutschen Infanterie im ersten Weltkrieg.* In: *Feldgrau – Mitteilungen einer Arbeitsgemeinschaft.* Lehrte (1954, 1956)

Kling, H.: *Formationen der Feldartillerie 1914 – 1918.* In: *Feldgrau – Mitteilungen einer Arbeitsgemeinschaft.* Lehrte (1956/1957)

Kling, H.: *Zur Formationsgeschichte der Infanterie 1914 – 1918. Die bei der Mobilmachung aufgestellten Landwehr-Inf.-Regimenter.* In: *Zeitschrift fuer Heereskunde* (1960)

Kling, H.: *Zur Formationsgeschichte der Infanterie 1914/18. Die bei der Mobilmachung aufgestellten Reserve-Regimenter.* In: *Zeitschrift fuer Heereskunde* (1958)

Kling, H.: *Zur Formationsgeschichte der Infanterie 1914/18. Die Brigade-Ersatzbataillone.* In: *Zeitschrift fuer Heereskunde* (1957)

Kling, H.: *Zur Geschichte der Neuformationen der Infanterie 1914/18 (Bayern).* In: *Zeitschrift fuer Heereskunde* (1961)

Priesdorff, K. v. ed.: *Soldatisches Führertum, Bd. 1 – 10*, Hamburg, Hanseatische Verlags-Anstalt, 1937 – 1942.

Reichsarchiv ed.: *Der Weltkrieg 1914 bis 1918, Das Marnedrama 1914*, Oldenburg, Druck und Verlag Gerhard Stalling, 1926.

Reichsarchiv ed.: *Der Weltkrieg 1914 bis 1918, Die Eroberung von Novo Georgiewsk*, Oldenburg, Druck und Verlag Gerhard Stalling, 1925.

Reichsarchiv ed.: *Der Weltkrieg 1914 bis 1918, Die Schlacht bei St. Quentin 1914*, Berlin, Druck und Verlag Gerhard Stalling, 1925.

Reichsarchiv ed.: *Der Weltkrieg 1914 bis 1918, Die Schlacht vor Paris, Das Marnedrama 1914*, Berlin, Druck und Verlag Gerhard Stalling, 1925.

Reichsarchiv ed.: *Ildirim, Deutsche Streiter auf heiligem Boden*, Druck und Verlag Gerhard Stalling, Berlin, 1924.

Reichsarchiv ed.: *Somme Nord, Die Brennpunkte der Schlacht im Juli 1916*, Berlin, Druck und Verlag Gerhard Stalling, 1927.

Roehrs, M.: *Matrosen-Regiment 3*. Oldenburg, 1921

Sproesser: Deutschlands Heerführer. Leipzig, Hirth Verlag, 1895

Stein, H. R. v.: *Die ganz hohen "Hausnummern"*. In: *Zeitschrift fuer Heereskunde* (1973)

Stein, H. R. v.: *Die Landsturm-Infanterie-Regimenter*. In: *Zeitschrift fuer Heereskunde* (1974)

Stein, H. R. v.: *Die Minenwerfer-Formationen 1914 – 1918*. In: *Zeitschrift fuer Heereskunde* (1959/1960)

Stein, H. R. v.: *Die Organisation der deutschen Kavallerie 1914 – 1918*. In: *Feldgrau – Mitteilungen einer Arbeitsgemeinschaft*. Lehrte (1959)

Stein, H. R. v.: *Die Pioniere 1914 – 1918*. In: *Zeitschrift fuer Heereskunde* (1976)

Stein, H. R. v.: Die reitenden Abteilungen der Feldartillerieregimenter 1914 – 1918. In: Feldgrau (?)

Stein, H. R. v.: *Landwehr- und Landsturmeskadronen 1914 – 1918*. In: *Feldgrau – Mitteilungen einer Arbeitsgemeinschaft*. Lehrte (1961)

Stein, H. R. v.: *Reserve-Kavallerie-Regimenter, Reserve-Kavallerie-Abteilungen und weiter Kriegsformationen 1914 – 1918*. In: *Feldgrau – Mitteilungen einer Arbeitsgemeinschaft*. Lehrte, Special Volume No. 17

Stein, H. R. v.: *Reserve-Kavallerie-Regimenter, Reserve-Kavallerie-Abteilungen und weitere Kriegsformationen 1914 – 1918*. In: *Feldgrau – Mitteilungen einer Arbeitsgemeinschaft*. Lehrte (1962)

Stein, H. R. v.: *Truppenteile von deutschen Kavallerieverbaenden im ersten Weltkrieg*. In: Feldgrau (1963)

Tannen, Illing, R., Schuetz, A., Forstner, K. v.: *Das kaiserliche Marine-Infanterie-Regiment Nr. 3 in den Stuermen des Weltkrieges 1914 – 18*. Zeulenroda, 1936

Voigt, G.: *Deutschlands Heere bis 1918. Bd. 1 – 10*, Osnabrueck, Biblio-Verlag, 1980 – 1984

Wacker, P., Schirmer, F.: *Die preussischen Garde-Infanterie-Regimenter 1912 – 18*. In: *Feldgrau – Mitteilungen einer Arbeitsgemeinschaft*. Lehrte (1954), Special Volume

Wacker, P.: *Die deutschen Seebataillone im Rahmen des deutschen Marinekorps in Flandern von 1914 – 1918*. In: *Feldgrau – Mitteilungen einer Arbeitsgemeinschaft*. Lehrte (1953)

COMPARATIVE PERFORMANCE OF GERMAN AND ALLIED ANTI-TANK WEAPONS IN WWII: Edited by G.F.Nafziger, 47 pgs, $12.95 plus S&H.

THE GERMAN TANK PLATOON IN WWII: ITS TRAINING AND EMPLOYMENT IN BATTLE: Edited By G.F.Nafziger. 114 pgs, 179 illustrations, $19.95.

GERMAN PANZER TACTICS IN WORLD WAR II, Combat Tactics of German Armored Units from Section to Regiment: By Charles C. Sharp, 101 pages, numerous illustrations, $19.95.

GERMAN SQUAD TACTICS WORLD WAR II, By Matthew Gajkowski. 106 pages, numerous illustrations, , $19.95.

SOVIET TACTICAL DOCTRINE IN WWII, 146 pgs, numerous illustrations, $19.95.

SOVIET ARMORED TACTICS IN WWII, THE TACTICS OF THE ARMORED UNITS OF THE RED ARMY FROM INDIVIDUAL VEHICLES TO BATTALIONS ACCORDING TO THE COMBAT REGULATIONS OF FEBRUARY 1944: By Charles C. Sharp, 18 diagrams, $19.95.

SOVIET INFANTRY TACTICS IN WORLD WAR II: Red Army Infantry Tactics from Squad to Rifle Company from the Combat Regulations, By Charles C. Sharp, 122 pages, numerous illustrations, $19.95.

WORLD WAR II CAMPAIGN AND OTHER STUDIES

THE WESTWALL (SIEGFRIED LINE) 1938-1945, By Jean-Denis Lepage, 94 pages, heavily illustrated, $19.95.

THE AFRIKA KORPS, AN ORGANIZATIONAL HISTORY 1941-1943, By G.F.Nafziger, 128 pages, many orders of battle, $19.95.

LESSONS LEARNED By THE AMERICAN ARMY DURING THE TUNISIAN CAMPAIGN: reprint of 1943 original government publication. 32 pages. Single volume for $12.95.

THE BERLIN DIRECTION: APRIL-MAY 1945, AN EXTRACTION OF THE OFFICIAL SOVIET ARMY ORDER OF BATTLE OF THE BERLIN STRATEGIC OFFENSIVE OPERATION APRIL-MAY 1945 by Craig Crofoot, 104 pages, $19.95.

THE SOVIET ORDER OF BATTLE: VOLUME 1, PART 1: THE NORTHERN THEATER OF OPERATIONS, 22 JUNE 1941 - 1 JULY 1942, by Craig Crofoot, 123 pages, $19.95.

THE AMERICAN ARMY IN WORLD WAR II: By George F. Nafziger. $19.95 per volume.

Vol 1 - 1st-40th Infantry Divisions (includes history of building of army from 1939-1945, plus tables of organization and organizational history of 1st-40th Divisions.),125 pages
Vol 2 - 41st-104th Infantry and Airborne Divisions (only organizational history of individual divisions), 119 pages
Vol 3 - Independent Armored, Mechanized Cavalry, Artillery, Anti-aircraft and Engineer Battalions, 82 pages
Vol 4 - Armored Divisions, 104 pages

THE BRITISH ARMIES OF THE SECOND WORLD WAR - AN ORGANIZATIONAL HISTORY: By David Hughes, James Groshot & Alan Philson
Vol 1 - Armoured and Cavalry Divisions - 123 pages
Vol 2 - Dominion Armored Divisions and British Infantry Divisions - 132 pages
Vol 3 - Territorial Army, War Raised Mountain and Training Infantry Divisions - 122 pages.
Vol 4 - Tank & Armoured Brigades, 79th Armoured Division, Armd Car Regiments,
African, Malayan & other Colonial Forces - 126 pages
Vol 5 - The Australian Army - 108 pages
Vol 6 - The Canadian Army - 106 pages
Vol 7 - New Zealand & South Africay Armies - 122 pages
Vol 8 - Indian Army, Part 1 - Indian Army in the West - 128 pages
Vol 9 - Indian Army, Part 2 - Indian Army in the East 1939-1943- 131 pages
Sup 1 - OB for the British Army in the early part of the war.
Sup 2 - OB for the British Army 1941-1942 - 109 pages
Sup 3 - OB for the British Army 1942-1944 - 97 pages
Sup 4 - OB for the British Army 1944-1945 - 111 pages

RUMANIAN ORDER OF BATTLE IN WORLD WAR II, By G.F.Nafziger, 90 pages, $19.95.

BULGARIAN ORDER OF BATTLE IN WORLD WAR II: By G.F.Nafziger, 75 pages, $19.95.

THE CHINESE ARMY, THE GROWTH AND ORGANIZATION OF THE CHINESE ARMY FROM 1895 TO 1945: By George F. Nafziger, 100 pages, $19.95.

ITALIAN ORDER OF BATTLE IN WORLD WAR II, By G.F.Nafziger. $19.95/Volume.
Vol 1 - Armored, Motorized, Airborne & Alpini Divisions, 108 pages.
Vol 2 - Infantry Divisions, 95 pages.
Vol 3 - Black Shirt, Mountain, & Assault Divisions and the 1944 Italian Army in Allied Service, 78 pages.

JAPANESE ORDER OF BATTLE WWII: By John L. Underwood, Jr. $19.95/Volume.

Vol 1 - Infantry & Depot Divisions, 129 pages.
Vol 2 - Airborne, Raiding, Amphibious, Cavalry and Independent Brigades, 113 pages.
Vol 3 - Independent Battalions, Army Detachments, and Miscellany, 94 pages.
Vol 4 - Japanese Armored Units in WWII - 85 pages
Vol 5 - Japanese Armor in Manchuria - 108 pages

JAPANESE INFANTRY ORGANIZATION & TACTICS IN WWII : Edited By G.F.Nafziger. $19.95/volume

Vol 1 - Basic Infantry Organization & Discussion of Overall Tactical System, including Training Cycle and Night Vision Techniques. - 135 pages
Vol 2 - Extracts from Japanese Tactical Manuals & US Intelligence Observations on Japanese Defensive Cave Tactics - 138 pages
Vol 3 - Historical Examples of Japanese Night Attacks - 103 pages

GERMAN ORDER OF BATTLE IN WWII, WAFFEN SS, FALLSCHIRMJAEGERS & FOREIGN TROOPS IN GERMAN SERVICE, By G.F.Nafziger, 394 pages $60.

This is a revised reissue of the third volume of the German Order of Battle series published several years ago. This particular volume covers every major formation in the Waffen SS, all the major non-flying Luftwaffe field formations, notably the 22nd Air Landing Division and all the Fallschirmjaegers, the Naval Divisions, the Slovak, Serbian, Italian and Russian volunteer units that were part of the Wehrmacht. Complete tables of organization that include numbers of heavy weapons, are included. Copies of the first edition are currently selling used for over $250.

OTHER WORLD WAR II

ROMMEL AND GUDERIAN AGAINST THE BELGIAN CHASSEURS ARDENNAIS, The Combats at Chabrehez and Bodange, 10 May 1940, Translated By G.F.Nafziger, 123 pages, 3 maps, $19.95.

CASE WHITE, THE GERMAN ARMY IN THE POLISH CAMPAIGN 1939, By William Russ, 271 pages, extensive appencies with complete orders of battle and 92 maps. $25.

TANK DESTROYERS IN COMBAT, Edited by G.F.Nafziger, 49 pgs, numerous illustrations, $12.95.

NAPOLEONIC TACTICS

NAPOLEON'S SYSTEM OF WAR, By General H. Camon, Translated & annotated By G.F.Nafziger.

A TREATISE UPON THE REGULATIONS OF THE FRENCH INFANTRY, By Général de brigade Meunier. Originally published 1809, 60 pages, numerous illustrations, $19.95.

TACTICS OF THE RUSSIAN ARMY IN THE NAPOLEONIC WARS, By Alexander & Yurii Zhmodikov, $19.95/Volume.

- Vol I - Russian Army in the 18th Century & Tactical Response to the French Revolution, 1801-1809 - 117 pages
- Vol 2 - Tactical Changes in Response to Napoleon - 1810-1814 - 125 pages

NAPOLEONIC WARS AND WARS OF THE FRENCH REVOLUTION

RUSSO-TURKISH WAR OF 1806-1812, Translated By A.Mikaberidze, 2 volumes, 149 & 145 pages, numerous battle maps, $25.00/Volume.

RUSSO-SWEDISH WAR OF 1808-1809, Translated By A.Mikaberidze, 1st volume, 182 pages, 27 maps & 17 black & white illustrations, $25.00.

AUSTERLITZ: A RUSSIAN ACCOUNT OF THE 1805 CAMPAIGN, By Lt. General Mikhailovski-Danilevski, Translated and Annotated by G.F.Nafziger, French translation published 1841, 9 maps, 139 pages $25.00

A CONTEMPORARY ACCOUNT OF THE 1796 CAMPAIGN IN GERMANY AND ITALY, by General Thomas Graham, Baryon Lynedoch, Edited & annotated By G.F.Nafziger, originally published 1800, 141 pages. $19.95.

A CONTEMPORARY ACCOUNT OF THE 1797 CAMPAIGN IN GERMANY AND ITALY, by General Thomas Graham, Baryon Lynedoch, Edited & annotated By G.F.Nafziger, originally published 1801. 117 pages. $19.95.

A CONTEMPORARY ACCOUNT OF THE 1799 CAMPAIGN IN GERMANY AND SWITZERLAND, by General Thomas Graham, Baryon Lynedoch, Edited & annotated By G.F.Nafziger, Originally published 1801, 140 pages, $19.95.

A CONTEMPORARY ACCOUNT OF THE 1799 CAMPAIGN IN ITALY, by General Thomas Graham, Baryon Lynedoch, Edited & annotated By G.F.Nafziger, Originally published 1801, 125 pages, 3 maps, $19.95.

A CONTEMPORARY ACCOUNT OF THE 1799 CAMPAIGN IN HOLLAND, by General Thomas Graham, Baryon Lynedoch, Edited & annotated By G.F.Nafziger, Originally published 1801, 171 pages, 3 maps, $25.00.

AN ENGLISH GENERAL IN THE ARMY OF REVOLUTIONARY FRANCE 1792, By General John Money, Annotated By George F. Nafziger, originally published 1794. 84 pages, $19.95.

MEMOIRS TO SERVE THE HISTORY OF THE 1796 CAMPAIGN IN GERMANY, By Général de division Jourdan, Edited & Annotated By G.F.Nafziger, Originally published 1818, 126 pages $19.95.

ARCHDUKE CHARLES' 1796 CAMPAIGN IN GERMANY, By Archduke Charles, Translated by G.F. Nafziger, French translation published 1840, 222 pgs, $25.00.

THE FIRST PHASE OF NAPOLEON'S 1796 CAMPAIGN, Translated & annotated By George F. Nafziger, originially published 1905. 103 pages, $25.00.

A STRATEGIC ANALYSIS OF THE HOHENLINDEN AND MARENGO CAMPAIGNS, by Heinrich von Blow, 79 pages. Originally published under the title *Histoire de la campagne de 1800, en Allemagne et Italie en 1804*, reprinted in French translation, with commentary by Napoleon, in 1831 .

HISTORY OF THE 1800 CAMPAIGN, By the Duke de Valmy, translated by G. F. Nafziger, 115 pages, $19.95.

THE EYLAU-FRIEDLAND CAMPAIGN OF 1806-7, By G.F.Nafziger, 103 pages, 14 maps and illustrations, $19.95.

THE LEGION DU NORD 1806-1808, Memoire of Major Coqueugniot, by Major Lazare Claude Coqueugniot, Translated & Annotated by G.F.Nafziger, 67 pages.

BRUNE'S 1807 CAMPAIGN IN SWEDISH POMERANIA, by Chevalier Vigier de Saint-Junien, translated and compiled By G.F.Nafziger, Originally published 1817. 73 pages, $19.95.

OPERATIONS OF THE POLISH ARMY DURING THE 1809 CAMPAIGN IN POLAND, Translated By G.F.Nafziger, 175 pages, 1 map. Originally published in 1841. $25.00.

OPERATIONS OF THE 7TH POLISH LIGHT CAVALRY DIVISION IN THE LEIPZIG CAMPAIGN, Translated by G. F. Nafziger, 35 pages, $12.95.

An anonymous work published in 1814. The author was a Polish engineering officer who apparently accompanied the division. It provides a day-by-day account of the actions of the 7th Cavalry Division, with very detailed accounts of its actions at Wachau, during the Battle of Leipzig.

HISTORICAL ACCOUNT OF THE MILITARY OPERATIONS OF THE ARMY OF ITALY IN 1813 AND 1814, By Général de division Comte de Vignolle, translated By G.F.Nafziger, 93 pages, 1 map, $19.95.

HISTORICAL DICTIONARY OF THE FRENCH REVOLUTION AND TERROR, By George F. Nafziger. 103 pages, index & bibliography, $19.95.

WARS OF THE FRENCH REVOLUTION, By the Society of Soldiers and Men of Letters. Originally published in 1818 under the title *Victoires, conquêtes, désastres, revers et guerres civiles des Française de 1792 - 1815*, in 1818 and reprinted in an edited version in 1900.

WARS OF THE FRENCH REVOLUTION, By the Society of Soldiers and Men of Letters, translated By George F. Nafziger,

Volume 1, 28 April 1792-27 August 1793, 153 pages, 9 maps.
Volume 2, 28 August 1793-26 May 1794, 148 pages, 7 maps.
Volume 3, 27 May 1794-1 January 1795, 141 pages, 15 maps.
Volume 4, 1 January 1795-14 October 1795, 177 pages, 5 maps.
Volume 5, 15 October 1795-18 June 1796, 155 pages, 5 maps.
Volume 6, 19 June – 20 August 1796, 167 pages, 5 maps.

WARS OF THE REVOLUTION, By Arthur Chuquet, translated and annotated by W.D.Peterson. This is a series of twelve books written by the famous French author, A. Chuquet. Each work has a separate title and is listed below. An 11 volume series originally published 1910. $25 per volume.

Volume 1 - **THE FIRST PRUSSIAN INVASION (11 AUGUST-2 SEPTEMBER 1792),** 177 pages
Volume 2 - **VALMY,** 164 pages
Volume 3 - **BRUNSWICK'S RETREAT,** 171 pages
Volume 4 - **JEMAPPES,** 135 pages, 5 B&W drawings, 1 map & 8 photographs
Volume 5 – **NEERWINDEN AND THE DEFECTION OF DUMOURIEZ 1793**, 152 pages, 1 map, and 10 B&W Photographs
Volume 6 – **CUSTINE'S INVASION OF GERMANY, THE OPERATIONS OF THE ARMY OF THE RHINE, AUGUST 1792-APRIL 1793,** 170 pages, 5 maps, 20 photographs,
Volume 7 – **THE SIEGE OF MAINZ AND THE FRENCH OCCUPATION OF THE RHINELAND 1792-1793,** 179 pages, 19 maps and photographs.
Volume 8 - **WEISSEMBOURG,** 138 pages, 1 map and 25 photographs..
Volume 9 - **HOCHE AND THE BATTLE FOR ALSACE (1793-1794)**, 146 pages, 15 B&W drawings and 28 photographs.

THE CAMPAIGNS OF THE ROYALIST ARMY UNDER THE DUC DE CONDE DURING THE WARS OF THE FRENCH REVOLUTION, by Lt. General d'Ecquevilly, Translated by G.F.Nafziger, Volume 1, 138 pages, $25.00; Volume 2, 137 pages, $25.00p; Volume 3, 94 pages, $19.95.

This is the operational career of an obscure counter-revolutionary army where knights, barons, and dukes fought in the ranks as simple soldiers against the hated revolutionaries. The complete work covers their operations from 1792 to 1801, when the army was disbanded. The first volume covers its founding in 1791 through 1794. It discusses the operations of the Royalist army against the Weissembourg lines in very great detail, down to the actions of small groups of men. The second volume discusses its incorporation into the Russian Army and the third volume covers its operations in the Hohenlinden campaign, its reincorporation into the British Army, and its eventual disbandment/reorganization into the Chasseurs Britanniques.

HISTORICAL REVIEW OF THE EXPLOITS AND VICISSITUDES OF CAVALRY DURING THE WARS OF THE REVOLUTION AND THE EMPIRE UP TO THE ARMISTICE OF 4 JUNE 1813, by L.A.Dallwitz, translated & annotated by G.F. Nafziger. Volume 1 - 1792-1807, 142 pages, Volume 2 - 1808-1813, 129 pgs. $25.00 per volume.

The original work by Dallwitz was translated by Unger into French and cut down to cover only the period from 1792 to 1813. In this work Dallwitz addresses various actions by cavalry during these wars in order to demonstrate the continuing utility of cavalry in warfare. In the annotation process, these actions were augmented by additional material and the analyzed to determine lessons that might be learned as well as to explore, in greater detail, the operations of cavalry during this evolutionary period.

NAPOLEONIC ARMY STUDIES

This series of booklets examines the army of each nation, providing a complete history of its organization, the raising, disbanding, and the modification of the organization of the army's units. Details are provided on uniforms and their changes. In addition, to a varying degree, the battle history of the army and its units is discussed. These booklets were designed to provide the wargame gamer what is needed to build any of these armies in miniature. They were prepared using archival materials, official documents ranging from regimental histories to governmental studies, and memoirs of members of these armies. Each volume has some black & white illustrations.

THE CRESCENT AMONG THE EAGLES OTTOMAN EMPIRE AND THE NAPOLEONIC WARS, By W.E.Johnson, 110 pages, $19.95.

THE PRUSSIAN ARMY DURING THE NAPOLEONIC WARS (1792-1815),
By George F. Nafziger. $19.95/ Volume.

Vol I - Infantry (1792-1815) 122 pages
Vol II - Guard & Landwehr (1792-1815) 70 pages
Vol III - Cavalry, Artillery, Technical Troops, & Train (1792-1815) 107 pages

THE BRITISH ARMY, ITS REGIMENTS AND BATTALIONS, 1793-1815, By Michael G. McKenna. $25 per volume

- Vol I - Administration & Organization, Civil Departments, Officers, Cavalry, Infantry and Fencible Units - 148 pg
- Vol II - Veteran & Garrison Units, Ordnance, Colonial Units, Foreign Units, KGL, Royal Marines, Militia & Volunteers, and the Honourable East India Company - 142 pgs

THE IMPERIAL RUSSIAN ARMY, 1763-1815, by George F. Nafziger & W. Worley, $19.95/Volume.

- Vol I - Infantry, Mixed Units, Legions, Marines, Engineers, Pioneers, Sappers & Opolochenie. 107 pages
- Vol II - Cavalry, Cossacks, Guard, and Artillery 103 pages

THE ARMIES OF THE KINGDOM OF BAVARIA AND THE GRAND DUCHY OF WÜRZBURG 1792-1815, By G.F.Nafziger, 102 pages, $19.95.

ROYAL, REPUBLICAN, IMPERIAL, A HISTORY OF THE FRENCH ARMY FROM 1792-1815, By George F. Nafziger. $19.95 per volume.

- Vol 1 - Infantry - History of Line Infantry (1792-1815), Internal & Tactical Organization; Revolutionary National Guard, Volunteers Fédérés, Compagnies Franches; and 1805 National Guard. 98 pages.
- Vol 2 - Infantry - National Guard after 1809; Garde de Paris, Gendarmerie, Police, & Colonial Regiments; Departmental Reserve Companies; and Infantry Uniforms. 103 pages.
- Vol 3 - Cavalry; Line, National Guard, Irregular, & Coastal Artillery, Artillery & Supply Train, and Balloon Companies. 127 pages.
- Vol 4 - Imperial Guard. 141 pages.
- Vol 5 - Foreign Regiments - Swiss, Spanish, King Joseph's Army, Croatians, Portuguese, Régiments d'étranger, Albanians, Hanoverian Legion, etc. 133 pages.

ARMIES OF THE GERMANY AND THE CONFEDERATION OF THE RHINE 1806-1815,

- Vol I. (Anhalt, Saxe-Gotha, Saxe-Hildburghausen, Saxe-Meinigen, Coburg, Frankfurt Primate, Hesse-Darmstadt, Nassau, Mecklenburg-Schwerin and Mecklenburg-Strelitz), By G.F.Nafziger, 104 pages, $19.95.
- Vol II. (Baden, Lippe, Reuss, Schwarzburg, and Waldeck), By G.F.Nafziger, 77 pages, $19.95.

THE ARMIES OF SPAIN AND PORTUGAL, 1808-1814, By G.F.Nafziger, 93 pages, $19.95.

THE WÜRTTEMBERG ARMY, 1792-1815, By G.F.Nafziger, 76 pages, $19.95.

NAPOLEON'S ENEMIES, THE ARMIES OF BRUNSWICK, HANOVER, HESSE-CASSEL, AND THE HANSEATIC CITIES, By G.F.Nafziger, 110 pages, $19.95.

THE ARMIES OF WESTPHALIA AND CLEVES-BERG, 1806-1815, By G.F.Nafziger, 80 pages, $19.95.

SEVEN YEARS WAR ARMY STUDIES

THE PRUSSIAN ARMY OF FRIEDRICH DER GROSSE, By G.F.Nafziger, $19.95 per volume.

Volume 1 - The Infantry - 92 pages
Volume 2 - The Saxons, Cavalry & Technical Troops - 92 pages

THE FRENCH ARMY OF LOUIS XV, In the War of the Austrian Succession and Seven Years War, By Lucien Moulliard, Translated and Annotated by G.F.Nafziger, Originally published in 1882. $50, 152 pages. 57 pages of color plates.

THE NEW USE OF ARTILLERY IN FIELD WARS: NECESSARY KNOWLEDGE, By Chevalier du Teil, translated by Charles Shallcross, 74 pages, some illustrations, $19.95. Originally published in 1778.

OTHER ERAS

NOTES ON INFANTRY TACTICS IN KOREA By S.L.A.Marshall and appendix by B.H.Liddell Hart. 102 pages, $19.95.

COMMENTARY ON INFANTRY OPERATIONS AND WEAPONS USAGE IN KOREA (WINTER OF 1950-51) By S.L.A.Marshall. 124 pages, $19.95.

THE AMERICAN ARMY IN WORLD WAR I, By G.F.Nafziger, 110 pages, $19.95.

THE RUSSIAN ARMY IN WORLD WAR I, By G.F.Nafziger, 107 pages, $19.95.

GERMAN ARMIES' ESTABLISHMENTS 1914/1918, By Dirk Rottgardt, $19.95/Volume. This series of works provides the detailed TO&Es for every formation of the German Armies in WWI.

Volume 1 - Higher Staffs, Infantry Regiments - 92 pgs
Volume 2 - Reserve and other Infantry Units - 102 pgs
Volume 3 - Cavalry and Supply Units - 84 pgs
Volume 4 - German Troops in the Orient - 84 pgs
Volume 5 - Engineers, Railway and Motor Transport - 121pgs
Volume 6 - Signals and Air Forces - 130 pgs

GERMAN NAVAL INFANTRY IN WORLD WAR I, By Dirk Rottgardt, 141 pages, $25.

GERMAN DIVISIONS OF WWI, By Dirk Rottgardt, $19.95 per volume

Volume 1 - Alpine Corps, Bavarian, Naval and Guard Divisions - 106 pgs
Volume 2 - 1st-13th Divisions – 105 pgs
Volume 3 - 14th-27th Divisions - 103 pgs
Volume 4 - 28th - 55th Divisions – 111 pgs
Volume 5 - 56th - 203rd Divisions – 117 pgs
Volume 6 - 204th-303rd Divisions, Independent Brigades, Asian Corps and Caucasian Training Command – 116 pgs
Volume 7 - Cavalry Divisions – 85 pgs

TO THE LAST CARTRIDGE, By Alan Curtis. Volume 1 - 172 pgs, Vol 2 - 135 pgs. $25.00 per volume.

This two-volume work is a comprehensive history of the War of the Pacific, between Chile, Peru and Bolivia. It lasted from 1879 to 1884 and was fought in the arid Atacama desert and the heights of the Andes, from the Equator to Cape Horn. It also included the only battle between ironclads after the Battle of Lissa in 1866. Many orders of battle and much uniform detail is included. Extensive use has been made of Spanish language sources. This is the only comprehensive military study of this war in English. Volume 2 to be released in late 2007.

THE BATTLE OF MALPLAQUET, By Maurice Sautai, Translated by G.F.Nafziger, 62 pgs, 8 maps, $19.95.

Written for the Historical Section of the French Army, this work explores the correspondence with various officers of the French Army and the Duke du Maine relating to the battle of Malplaquet. It is a very thorough study of the battle.

Orders of Battle from
The Nafziger Collection, Inc.
PO Box 1522
West Chester, OH 45069-1522

The Nafziger Collection is a mass of orders of battle covering the period from 1600 to 1945. Sources range from published works to actual archival documents, which represent the largest single source. Nearly all OB's break down to the regimental level. The availability of strength figures and artillery equipment varies from period to period.

The catalog comes in sections for your convenience. All of the catalog sections are available for $6.00. The cost of the individual sections are listed below. Prices of the OB's are $0.25 per page, with discounts on large orders and multi-paged OB's. Full details on ordering come with the catalogs. The catalogs are available free on-line at http://home.fuse.net/nafziger.

- 17th Century Catalog
  - 1600-1699, including
    - 30 Years War & English Civil War
    - Total 150 OB's
    - Catalog Price $0.50 - 2 pages
- 18th Century Catalog:
  - 1697-1792
    - Spanish Succession Great Northern War
    - Polish Successiòn, Austrian Succession
    - Seven Years War, Bavarian Succession
    - Total 480+ OB's
    - Catalog Price $1.00 - 8 pages
- French Revolution Catalog:
  - 1792-1799
    - Total 450+ OB's
    - Catalog Price $1.00 - 5 pages
- Napoleonic Wars Catalog
  - 1800-1815
    - Includes Russo-Swedish 1808-9 War, Russo-Turkish Wars, Continental Europe and Peninsular Wars.
    - Total 1,400+ OB's
    - Catalog Price $1.50 - 15 pages
- Miscellaneous American Wars Catalog:
  - 1775-1864
    - American Revolution, War of 1812, Mexican-American War, Franco-Mexican War. (Civil War has a separate catalog)
    - Total 400+ OB's
    - Catalog Price $1.00 - 5 pages
- American Civil War Catalog
  - 1861-1865
    - Total 1,300+ OB's
    - Catalog Price $1.50 - 11 pages
- Post Napoleonic Catalog
  - 1816-1899
    - Russo Turkish War 1829, French in Mor , e, Belgian Revolution, French in Algeria 1830, Polish Revolution 1832, Carlist War 1834, 1848 Revolutions, Roman Republic 1849, Crimea 1854, Italian Revolution 1859, Prusso-Danish War 1864, Prusso-Austrian War 1866, Franco-Prussian War 1870, Russo-Turkish War 1877
    - Total OB's - 200+
    - Catalog Price $0.50 - 3 pages
- British Colonial Catalog
  - 1861-1865
    - Northwest Indian Frontier, 2nd Afgan War, China, Boer, Zulu and others
    - Total 75+ OB's
    - Catalog Price $0.50 - 2 pages
- Naval Orders of Battle Catalog
  - 1688-1801 - 500+ OB's
  - WWI - 30 OB's
    - Catalog Price $1.00 - 8 pages
- World War I Catalog
  - 1905-6 Russo-Japanese War:
    - 10 OB's
  - 1914-1918 World War I
    - Eastern & Western Front
    - Russian, Austrian, German, Serbian, Romanian, American, British & French
    - Total 200 OB's and growing
    - Catalog Price $0.50 - 5 pages
- Spanish Civil War
  - 1936-1939 125 OBs
    - Catalog Price $0.50 - 3 pages
- World War II Catalog
  - 1939-1945
    - Naval, Pacific, Mediterranean, North Africa, Europe.
    - Finnish, Russian Front, Vichy French, USMC in the Pacific, lots of TO&E material on every army.
    - 2,400+ OB's (over 8,500 pages of data)
    - Catalog Price $3.00 - 41 pages